Workbook for Basic Mathematics for Electricity and Electronics

8th Edition

Bertrand B. Singer
Late Assistant Chairman
Mathematics Department
Samuel Gompers High School
New York, New York

Harry Forster
Electronics Department
Miami-Dade Community College
University of Miami
Miami, Florida

Mitchel E. Schultz
Electronics Department
Western Wisconsin Technical College
La Crosse, Wisconsin

 **Glencoe McGraw-Hill**

New York, New York   Columbus, Ohio   Woodland Hills, California   Peoria, Illinois

Cover photos: (l) John Paul Endress/The Stock Market, (c) John Lund/The Stock Market, (r) Doug Martin

Basic Mathematics for Electricity and Electronics ISBN 0-02-805022-3
Workbook for Basic Mathematics for Electricity and Electronics ISBN 0-02-805023-1
Instructor's Manual for Basic Mathematics for Electricity and Electronics ISBN 0-02-805024-X

Glencoe/McGraw-Hill

A Division of The McGraw·Hill Companies

Workbook for Basic Mathematics for Electricity and Electronics

Send all inquiries to:
Glencoe/McGraw-Hill
8787 Orion Place
Columbus, OH 43240

ISBN 0-02-805023-1

Printed in the United States of America

2 3 4 5 6 7 8 9   047   07 06 05 04 03 02 01 00

Contents

Preface

The fields of electricity and electronics are closely tied to the field of mathematics. Through the study of mathematics, you can not only learn to solve electrical and electronics problems but you can also gain a better understanding of the nature of electricity and electronics.

This workbook was written to supplement the text *Basic Mathematics for Electricity and Electronics*, Eighth Edition. Its focus is to provide practice and to reinforce the mathematics taught in the student edition. This workbook contains problems representing each job from the student edition. You will find both computational problems and real-life applications.

The workbook can be used for individual homework assignments or for in-class collaborative group assignments.

A special thank you goes to Walter L. Bartkiw. His contributions and dedication made this workbook possible.

Harry Forster, Jr.
Mitchel E. Schultz

Introduction to Electricity

JOB 1-1	Basic Theory of Electricity

Answer each of the following.

1. Describe an atom.

2. Identify the particles contained within an atom.

3. Describe the structure of an atom and draw the structure of a lithium atom.

4. Why is the valence electron of a copper atom considered to be a free electron?

5. Define the term *voltage* and list its unit of measure.

6. How many electrons are contained in 1 coulomb of electricity?

7. Explain what current is and list its unit of measure.

8. Define the term *resistance* and list its unit of measure.

9. Explain the difference between a conductor and an insulator.

10. Identify the symbols for voltage, current, and resistance.

JOB 1-2   Electrical Measurements and Circuits

Answer each of the following.

1. Explain how an ammeter must be connected in a circuit to measure the current flow. Should the resistance of an ammeter be low or high? Why?

2. Explain how a voltmeter must be connected to measure the voltage in a circuit.

3. Explain how an ohmmeter must be connected to measure resistance in a circuit. What precaution must be observed when measuring resistance in a circuit?

4. Explain the difference between a complete circuit and an open circuit.

5. Explain the purpose of a fuse.

6. What type of circuit is created when a fuse blows?

7. Under what conditions will a fuse blow?

8. Describe the operation of a delayed fuse.

9. Draw the circuit symbols for the following electrical devices: (*a*) battery, (*b*) fixed capacitor, (*c*) fuse, (*d*) ground, (*e*) iron core inductor, (*f*) fixed resistor, (*g*) switch, (*h*) transformer, (*i*) semiconductor diode, and (*j*) NPN transistor.

10. List the components that every electric circuit must contain.

11. Draw a circuit that contains a battery, fuse, switch, ammeter, variable resistor, fixed resistor and a lamp. Label the diagram completely.

Simple Electric Circuits

2

c h a p t e r

Fractions

Express the following fractions in lowest terms.

1. $\dfrac{3}{24}$ 2. $\dfrac{6}{16}$ 3. $\dfrac{3}{9}$ 4. $\dfrac{12}{15}$ 5. $\dfrac{50}{75}$

6. $\dfrac{6}{8}$ 7. $\dfrac{9}{16}$ 8. $\dfrac{14}{20}$ 9. $\dfrac{21}{84}$ 10. $\dfrac{2}{22}$

Change each of the following mixed numbers to improper fractions.

11. $3\dfrac{3}{8}$ 12. $7\dfrac{4}{9}$ 13. $13\dfrac{3}{4}$ 14. $5\dfrac{2}{5}$ 15. $3\dfrac{1}{4}$

16. $2\dfrac{2}{6}$ 17. $6\dfrac{1}{3}$ 18. $1\dfrac{1}{2}$ 19. $30\dfrac{3}{9}$ 20. $14\dfrac{3}{7}$

Change the following improper fractions to mixed or whole numbers.

21. $\dfrac{33}{6}$ 22. $\dfrac{6}{4}$ 23. $\dfrac{9}{2}$ 24. $\dfrac{22}{3}$ 25. $\dfrac{60}{4}$

26. $\dfrac{75}{9}$ 27. $\dfrac{58}{8}$ 28. $\dfrac{55}{6}$ 29. $\dfrac{189}{12}$ 30. $\dfrac{35}{16}$

Multiply the following fractions and reduce to lowest terms.

31. $\dfrac{1}{2} \times \dfrac{3}{8}$

32. $\dfrac{1}{4} \times \dfrac{5}{10}$

33. $16 \times \dfrac{1}{2}$

34. $25 \times \dfrac{3}{4}$

35. $\dfrac{1}{2} \times \dfrac{1}{2}$

36. $15 \times \dfrac{3}{5}$

37. $\dfrac{3}{9} \times \dfrac{2}{6}$

38. $3 \times \dfrac{4}{6}$

39. $3 \times \dfrac{1}{3}$

40. $4 \times \dfrac{2}{9}$

Multiply the following mixed numbers.

41. $3\dfrac{1}{3} \times 2\dfrac{4}{6}$

42. $5\dfrac{2}{3} \times 3\dfrac{1}{2}$

43. $9\dfrac{2}{4} \times 4\dfrac{2}{3}$

44. $6\dfrac{1}{8} \times 3\dfrac{1}{4}$

45. $1\dfrac{3}{4} \times 1\dfrac{3}{4}$

46. $2\dfrac{1}{4} \times 3\dfrac{1}{8}$

47. $1\dfrac{4}{5} \times 2\dfrac{1}{8}$

48. $1\dfrac{1}{5} \times 2\dfrac{2}{5}$

49. $6\dfrac{2}{3} \times 2\dfrac{1}{4}$

50. $1\dfrac{3}{4} \times 2\dfrac{1}{4}$

Answer each of the following.

51. A resistor in a series circuit drops ⁹⁄₂₇ of the voltage applied to the circuit. Express the fraction in lowest terms.

52. An audio amplifier converts ⁹⁄₂₇ of its input power into sound output power. Express the fraction in lowest terms.

53. A No. 14 gage wire is carrying ¹²⁄₂₀ of its rated ampacity. Express the fraction in lowest terms.

54. The turns ratio of a voltage step-up transformer is $^{120}/_{720}$. Express the fraction in lowest terms.

55. The ratio of base current to collector current in a transistor equals $^{18}/_{1260}$. Express the fraction in lowest terms.

56. A 75-W light bulb draws $^{5}/_{8}$ A of current from the 120-V power line. How much current will three 75-W light bulbs draw?

57. The current in one branch of a parallel circuit is $^{3}/_{8}$ of the total current. How much is this branch current if the total current is 24 A?

58. The voltage drop across one resistor in a series circuit is $^{2}/_{7}$ of the total voltage. If the total voltage is 21 V, how much voltage is dropped across the resistor?

59. A 200-ft roll of No. 12 gage wire is used to wire the basement of a house. If ⅓ of the roll is used to wire the furnace room, how many feet of wire are left on the roll?

60. The current carried by a fuse is ⅛ its current rating, which is 30 A. How much current is the fuse carrying?

61. If an electrician makes an hourly rate of $15.50 per hour, how much does he earn if he works 3⅓ hours?

62. An electric heater uses 1½ kWh of energy in one hour. How much energy does the heater use in 3¾ hr?

63. Each section of a radio tower has a height of 9¾ ft. What is the total height of the tower if 6½ sections are above ground?

64. A certain fuse can carry a current 2½ times its rated value for 1 s. If the fuse has a current rating of ¼ A, how much current can the fuse safely carry for 1 s?

65. A ½-W, 1800-Ω resistor cannot safely handle a voltage more than 30 V without burning up. If the voltage across the resistor is 2⅓ times more than 30 V, how much voltage is across the resistor?

NAME ______________________ DATE ____________

JOB 2-3 Ohm's Law

Determine the voltage *V* for each of the following problems.

1. $I = \frac{1}{5}$ A, $R = 1000\ \Omega$

2. $I = \frac{1}{10}$ A, $R = 250\ \Omega$

3. $I = \frac{4}{500}$ A, $R = 2000\ \Omega$

4. $I = \frac{1}{25}$ A, $R = 300\ \Omega$

5. $I = \frac{2}{500}$ A, $R = 6000\ \Omega$

6. $I = 0.02$ A, $R = 470\ \Omega$

7. $I = 0.015$ A, $R = 1800\ \Omega$

8. $I = 0.003$ A, $R = 680\ \Omega$

9. $I = 1.05$ A, $R = 400\ \Omega$

10. $I = 0.75$ A, $R = 5\ \Omega$

For Figs. 2-1 to 2-10, solve for the voltage *V*.

11.

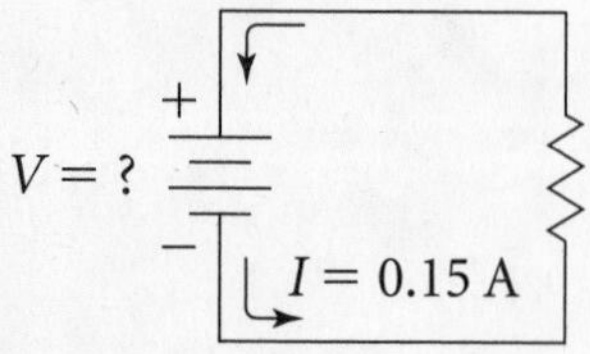

+
V = ?
−
R = 250 Ω
I = 0.15 A

Figure 2-1

12.

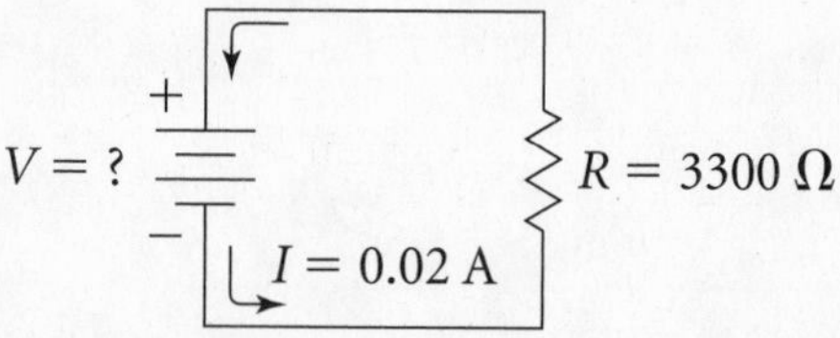

+
V = ?
−
R = 3300 Ω
I = 0.02 A

Figure 2-2

13.

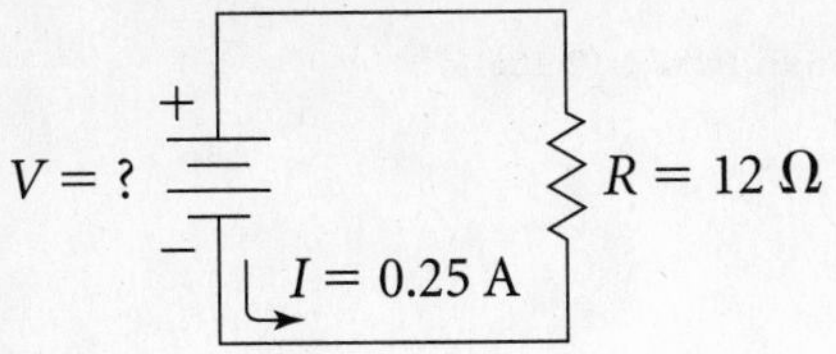
+
V = ?
–
R = 12 Ω
I = 0.25 A

Figure 2-3

14.

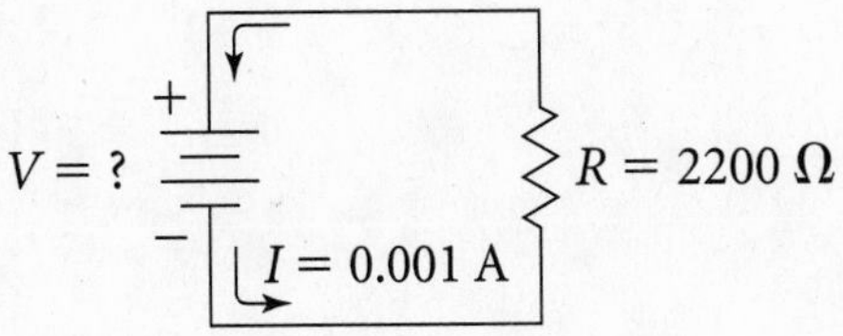
+
V = ?
–
R = 2200 Ω
I = 0.001 A

Figure 2-4

15.

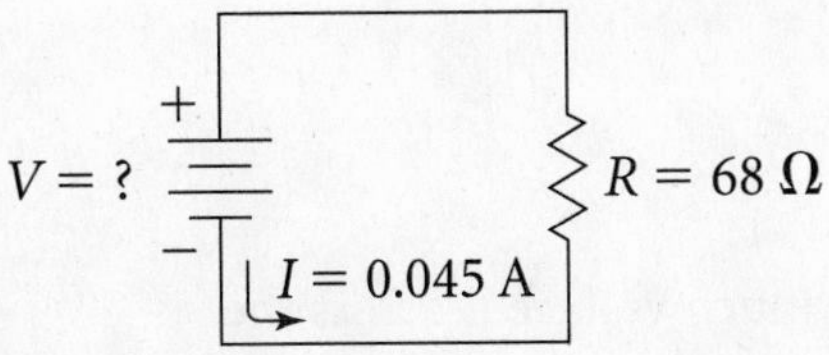
+
V = ?
–
R = 68 Ω
I = 0.045 A

Figure 2-5

16.

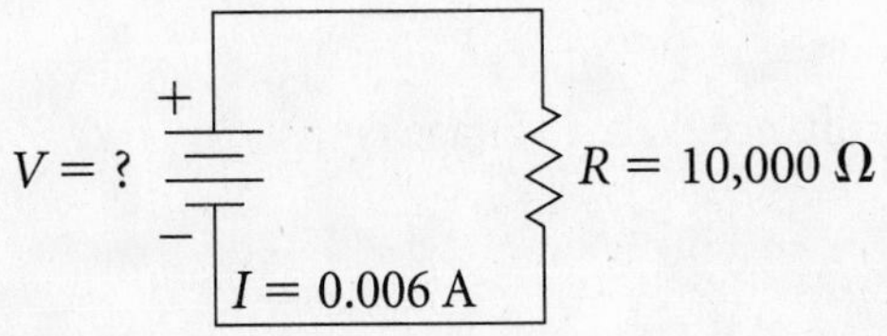
+
V = ?
–
R = 10,000 Ω
I = 0.006 A

Figure 2-6

17.

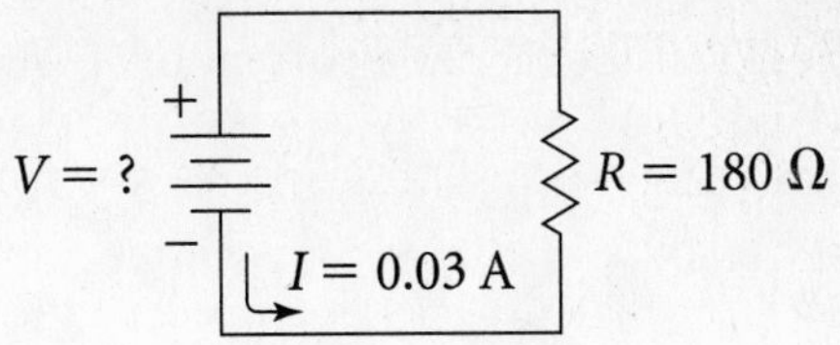
+
V = ?
–
R = 180 Ω
I = 0.03 A

Figure 2-7

18.

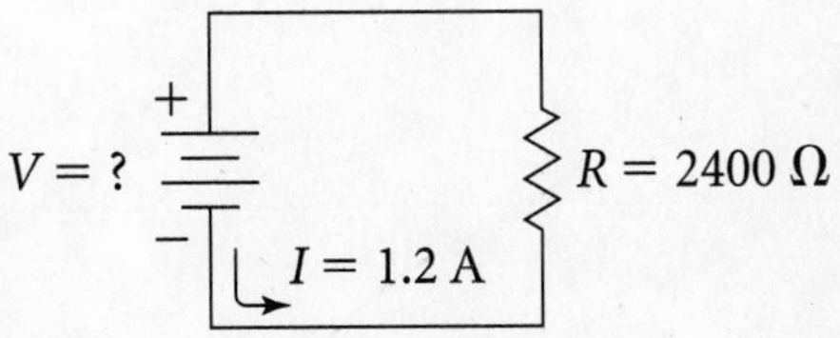
+
V = ?
–
R = 2400 Ω
I = 1.2 A

Figure 2-8

19.

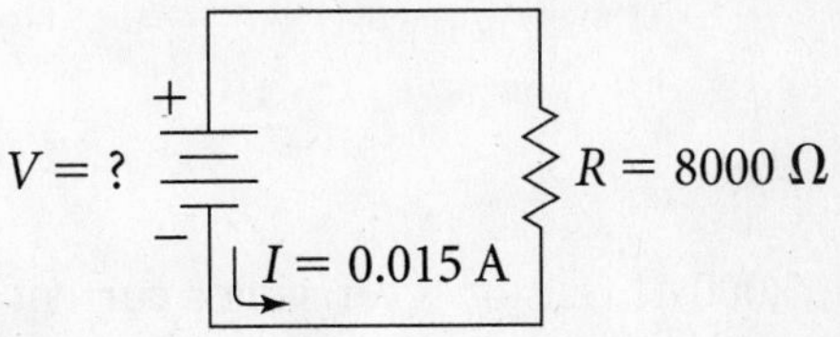
+
V = ?
–
R = 8000 Ω
I = 0.015 A

Figure 2-9

20.

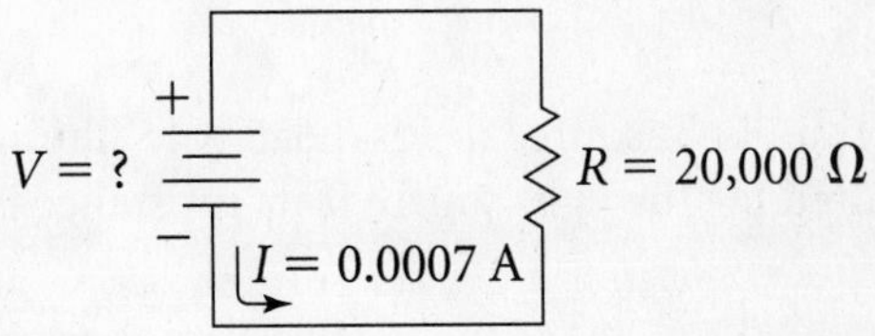
+
V = ?
–
R = 20,000 Ω
I = 0.0007 A

Figure 2-10

Solve each of the following.

21. A 1000-Ω resistor carries a current of $\frac{1}{100}$ A. How much voltage is across the resistor?

22. The resistance of a 1000-ft copper wire is 1.05 Ω. How much voltage is dropped across the wire if the current is 20 A?

23. How much voltage will be dropped across a 240-Ω relay coil that carries a current of 0.05 A?

24. A 2,200,000-Ω resistor is carrying a current of 0.00005 A. How much voltage is across the resistor?

25. A certain light source needs 0.625 A to illuminate with full brilliance. What voltage is required for the light source if its resistance is 192 Ω?

JOBS 2-4 and 2-5 | Decimals

Write the following fractions as decimals.

1. $\dfrac{40}{1000}$

2. $\dfrac{25}{100}$

3. $\dfrac{7}{10}$

4. $\dfrac{66}{1000}$

5. $\dfrac{20}{100}$

6. $\dfrac{15}{100}$

7. $\dfrac{3}{10}$

8. $\dfrac{2}{100}$

9. $\dfrac{7}{1000}$

10. $\dfrac{180}{1000}$

Arrange the following decimals in order beginning with the largest.

11. 0.015, 0.15, 0.45

12. 0.9, 0.04, 0.86

13. 0.01, 0.1, 0.099

14. 0.27, 0.269, 0.099

15. 0.75, 0.080, 0.8

Change the following mixed numbers to decimals.

16. $5\dfrac{30}{100}$ 17. $21\dfrac{77}{1000}$ 18. $12\dfrac{92}{1000}$ 19. $94\dfrac{13}{100}$ 20. $100\dfrac{6}{1000}$

21. $3\dfrac{10}{100}$ 22. $19\dfrac{6}{10}$ 23. $25\dfrac{30}{100}$ 24. $43\dfrac{2}{100}$ 25. $15\dfrac{2}{10}$

Change the following fractions into equivalent decimals.

26. $\dfrac{7}{8}$ 27. $\dfrac{6}{7}$ 28. $\dfrac{12}{13}$ 29. $\dfrac{1}{4}$ 30. $\dfrac{2}{5}$

31. $\dfrac{6}{16}$ 32. $\dfrac{3}{8}$ 33. $\dfrac{2}{4}$ 34. $\dfrac{1}{6}$ 35. $\dfrac{3}{4}$

Use Table 2-3 in your textbook to find the decimal equivalent of each of the following fractions and mixed numbers.

36. $\dfrac{7}{16}$ 37. $3\dfrac{15}{16}$ 38. $\dfrac{19}{32}$ 39. $4\dfrac{1}{4}$ 40. $8\dfrac{29}{64}$

Use Table 2-3 in your textbook to find the fraction nearest in value to each of the following decimals.

41. 0.49   42. 0.815   43. 0.297   44. 0.985   45. 0.94

Using Table 2-3 in your textbook, write each of the following decimals as a fraction or mixed number.

46. 0.6875   47. 2.468   48. 7.5625   49. 5.3   50. 0.344

Answer each of the following (refer to Table 2-3 in your textbook).

51. A 75-W light bulb draws a current of ⅝ A, and a CD player draws a current of 0.6 A. Which item draws more current?

52. A certain chip resistor is ⅛ in. long by ¹⁄₁₆ in. wide. Find the equivalent decimal value for each of these fractions.

53. Which is longer, a $1\frac{31}{64}$-in. screw or a 1.4375-in. screw?

54. A copper wire has a diameter of 0.15625 in., and an aluminum wire has a diameter of $\frac{5}{32}$ in. Which wire has the larger diameter?

55. A steel wire has a resistance of $6\frac{3}{8}$ Ω, and a silver wire has a resistance of 6.40625 Ω. Which wire has more resistance?

JOB 2-6 | Introduction to the Metric System of Measurement

Multiply the following numbers.

1. 1.63×100

2. 40.2×1000

3. 0.0085×10

4. $226 \times 10,000$

5. 1.05×100

6. 4.7×10

7. 0.0225×100

8. $1.75 \times 100,000$

9. 336.5×100

10. 0.005×100

11. 0.00016×100

12. $0.025 \times 10,000$

13. 263×10

14. 0.0000415×1000

15. $24 \times 10,000$

Divide the following numbers.

16. $256 \div 100$

17. $4700 \div 1000$

18. $850 \div 10$

19. $24.7 \div 10,000$

20. $510 \div 100$

21. $6 \div 100$

22. $1.2 \div 100,000$

23. $21 \div 100$

24. $0.01 \div 100$

25. $0.5 \div 100$

26. $100 \div 10,000$

27. $0.1 \div 100$

28. $5.6 \div 10,000$

29. $2.5 \div 100$

30. $105 \div 1000$

Perform the following conversions.

31. 5 cm to decimeters

32. 15 dm to centimeters

33. 1000 m to kilometers

34. 168 cm to kilometers

35. 10 dam to centimeters

36. 500 mm to centimeters

37. 25 cm to millimeters

38. 1405 m to kilometers

39. 0.5 km to millimeters

40. 450,000 mm to meters

41. 31 hm to centimeters

42. 500 dm to meters

43. 55 m to dekameters

44. 100 km to meters

45. 20,000 cm to meters

Answer each of the following.

46. The velocity of a radio wave in free space is 300,000,000 meters per second (300,000,000 m/s). What is the velocity in hectometers (hm) per second?

47. The wavelength of a radio wave is 20 m. What is the wavelength in centimeters?

48. The length of a resistor is 25 mm. What is its length in meters (m)?

49. A wire antenna has a length of 400 dm. What is its length in meters (m)?

50. A length of RG-8U coax is 2 km long. What is its length in mm?

JOB 2-7   Decimals

Perform the indicated operations.

1. $14.5 + 0.9 + 3.6$ 2. $5.03 + 7.0 + 0.111$ 3. $10.09 - 3.6$ 4. $16.1 - 2.9$

5. $4.22 + 3.18 + 4.02$ 6. $7.25 - 2.08$ 7. $13.65 - 1.88$ 8. $2.26 + 3.15 + 2.33 + 3.96$

9. $14.68 - 2\frac{1}{5}$ 10. $25 + 2\frac{1}{8} + 3.33 + 0.066 + 1.11$

Perform the indicated operations.

11. $3.9 \div 1.3$ 12. $5.6 \div 0.07$ 13. 3.35×17 14. 12.6×0.8

15. 15.4×0.06 16. $4.2 \div 1.05$ 17. 2.75×3 18. $6.08 \div 1.2$

19. $7.5 \div 1.5$ 20. 10.25×2.2

Answer each of the following.

21. A 1200-ft length of wire has a resistance of 15.6 Ω. How much is the resistance per foot of wire?

22. The current through an 11.5-Ω resistor is 0.3 A. How much voltage is dropped across the resistor?

23. A 24-Ω resistor has a voltage drop of 4.8 V. If the current $I = \dfrac{V}{R}$, how much current is flowing in the resistor?

24. A parallel circuit has the following branch currents: $I_1 = 1.003$ A, $I_2 = 4.32$ A, $I_3 = 0.007$ A, and $I_4 = 2.06$ A. Calculate the total current I_T if $I_T = I_1 + I_2 + I_3 + I_4$.

25. A parallel circuit with two branches has a total current I_T of 1.45 A. If $I_2 = 0.9$ A, find I_1 if $I_1 = I_T - I_2$.

Formulas 3

JOBS 3-1 and 3-2	Formulas

Write a formula for each of the rules stated. Use the letters and abbreviations in each problem to indicate each word.

1. The total resistance R_T of a parallel circuit equals the applied voltage V_A divided by the sum of the branch currents I_1 and I_2.

2. The total voltage V_T applied to a series circuit equals the current I multiplied by the sum of resistors R_1, R_2, and R_3.

3. The charge Q stored by a capacitor equals the product of the capacitance C and the voltage V.

4. In a current meter, the shunt current I_{SH} equals the difference between the total current I_T and the meter current I_M.

5. The current I carried by a series resistor R_S equals the difference between the source voltage V_S and the zener voltage V_Z divided by the resistance R_S.

6. An unknown resistor R_{unk} equals the product of resistors R and R_T divided by the difference between resistors R and R_T.

7. In a Wheatstone bridge, the unknown resistor R_x equals the standard resistor R_S multiplied by the quotient of R_1 divided by R_2.

8. The dc collector voltage V_C in a transistor circuit equals the difference between the collector supply voltage V_{CC} and the product of the collector current I_C and the collector resistance R_C.

9. The inductance L of an inductor equals the inductive reactance X_L divided by the product of 2, π, and the frequency f.

10. The total inductance L_T of two series inductors L_1 and L_2 with series-aiding mutual inductance L_M equals the sum of L_1, L_2, and twice the mutual inductance L_M.

Evaluate the following expressions.

11. $10 \div 5 + 2 \times 4$ 12. $24 \times 3 - 6 + 9 - 100 \div 4$ 13. $8(4 + 5) - 7$ 14. $13 - 4 \div 2 + 6 \times 12$

15. $116 - 3(9 + 3)$ 16. $14 + \dfrac{5}{20} - 48 \div 3(2 + 6)$ 17. $13(8 - 5) \div 3 + 7$

18. $\dfrac{2}{5} + 4 \div 10 + 2(0.3 \times 4)$ 19. $150 \div 3 + 2 - 4(11 - 3)$ 20. $5 \times 6 - 3(12 - 5)$

Solve the following problems using the formulas given in each problem.

21. Using the formula $I = \dfrac{P}{V}$, find the current I if P is 750 W and V is 120 V.

22. Using the formula $I_T = I_1 + I_2 + I_3 + I_4$, find the total current I_T in amperes if $I_1 = 1.4$ A, $I_2 = 0.8$ A, $I_3 = 1.1$ A, and $I_4 = 2.7$ A.

23. Using the formula $V = \dfrac{R}{R_T} \times V_T$, find V if $R = 1500\ \Omega$, $R_T = 5000\ \Omega$, and $V_T = 60$ V.

24. Using the formula $I = \dfrac{V}{R_1 + r_m}$, find I if $V = 10$ V, $r_m = 2000\ \Omega$, and $R_1 = 198{,}000\ \Omega$.

25. Using the formula $R_x = \dfrac{R R_T}{R - R_T}$, find R_x if R is $100\ \Omega$ and R_T is $60\ \Omega$.

26. In a series circuit the total resistance $R_T = R_1 + R_2 + R_3 +$ etc. What is the total resistance of a series circuit if R_1 is $20\ \Omega$, R_2 is $60\ \Omega$, and R_3 is $120\ \Omega$?

27. The internal resistance r_i of a battery can be calculated using the formula $r_i = \dfrac{V_{NL} - V_{FL}}{I_L}$.

What is the internal resistance of a battery if $V_{NL} = 9$ V, $V_{FL} = 8.4$ V, and $I_L = 0.15$ A?

28. The inductive reactance X_L of an inductor is calculated as $X_L = 6.28fL$. What is the inductive reactance of an inductor if f equals 1000 Hz and $L = 0.5$ H?

29. The capacitive reactance X_C of a capacitor is calculated using the formula $X_C = \dfrac{0.159}{fC}$. What is the capacitive reactance of a capacitor if f equals 500 Hz and $C = 0.0001$ F?

30. The dc collector-emitter voltage V_{CE} of a transistor circuit is calculated as $V_{CE} = V_{CC} - I_C(R_C + R_E)$. Calculate V_{CE} if $V_{CC} = 15$ V, $I_C = 0.006$ A, $R_C = 1000$ Ω, and $R_E = 250$ Ω.

JOB 3-3   Positive and Negative Numbers

Write the following quantities as signed numbers.

1. An increase of 20 dB (decibels)

2. A loss of 6 dB

3. 1000 ft below sea level

4. A reduction of $500 in savings

5. 10 A of current flowing to the right

6. If earth ground is considered to be 0 V, indicate a voltage that is 100 V above ground

7. A decrease in speed of 20 mph

8. A temperature of 30° below zero

9. 5 A of current flowing downward

10. A loss of 10 μC

Add the signed numbers in the following problems.

11. $(+25) + (+15)$

12. $(-25) + (-15)$

13. $(+6) + (-9)$

14. $(+90) + (-40)$

15. $(-45) + (-6)$

16. $(+9) + (+100)$

17. $(-5) + (+10)$

18. $(-21) + (-16)$

19. $(+18) + (-27)$

20. $(-13) + (-12)$

21. $(-6) + (-1)$

22. $(+6) + (-1)$

23. $(+75) + (+25)$

24. $(-80) + (+32)$

25. $(-10) + (+10)$

Multiply the following signed quantities.

26. $(-6) \times (-12)$

27. $(+13) \times (-3)$

28. $(+25) \times (+6P)$

29. $(+9) \times (+6)$

30. $(-7A) \times (+3)$

31. $(-16) \times (+4)$

32. $(-12) \times (-4)$

33. $(+18) \times (+2)$

34. $(-8) \times (-8)$

35. $(+15) \times (-7)$

Divide the following signed quantities.

36. $(+120) \div (+10)$ 37. $(-60) \div (+3)$ 38. $(+100) \div (-5)$ 39. $(-40) \div (-8)$

40. $(-75b) \div (-15)$ 41. $(+42g) \div (+6)$ 42. $(-6) \div (+12)$

43. $(+33) \div (-11)$ 44. $(-10) \div (-2)$ 45. $(-80) \div (+320)$

JOBS 3-4 and 3-5	Exponents

Solve the following problems.

1. Using the formula $A_P = A_V^2 \times \dfrac{R_{in}}{R_L}$, solve for A_P if $A_V = 150$ $R_{in} = 50 \ \Omega$, and

   $R_L = 5 \ k\Omega$.

2. Using the formula $A = \dfrac{\pi d^2}{4}$, find A if $\pi = 3.14$ and $d = 2$ inches (in.).

3. Using the formula $Z_P = a^2 \times Z_S$, find Z_P if $a = 4$ and $Z_S = 50 \ \Omega$.

4. Using the formula $Z_S = \dfrac{Z_P}{a^2}$, find Z_S if $Z_P = 1000 \ \Omega$ and $a = 3$.

5. Using the formula $A = \dfrac{\mu A}{10^6}$, change 5000 μA to amperes.

6. The power dissipated by a resistance can be calculated using the formula $P = I^2R$. Find P if $I = 0.1$ A and $R = 50\ \Omega$.

7. The power dissipated by a resistance can be calculated using the following formula:

$P = \dfrac{V^2}{R}$. Find P if $V = 20$ V and $R = 50\ \Omega$.

8. Using the formula $R = \dfrac{V^2}{P}$, find R if $V = 120$ V and $P = 100$ W.

9. Using the formula $R = \dfrac{P}{I^2}$, find R if $P = 60$ W and $I = 0.5$ A.

10. Using the formula $P_T = P_C\left(1 + \dfrac{m^2}{2}\right)$, find P_T if $P_C = 1500$ W and $m = 0.6$.

JOB 3-6	Powers of 10

Multiply the following numbers.

1. 0.033×10^4 2. 0.00047×10^6 3. 15.35×1000 4. $470,000 \times 10^1$ 5. 0.00068×10^3

6. 0.018×10^5 7. 0.0027×10^2 8. 3.46×10^3 9. 0.0000001×10^{10} 10. 0.0039×10^4

Divide the following numbers.

11. $25,000 \div 10^3$ 12. $476 \div 10^4$ 13. $667 \div 10^1$ 14. $75 \div 10^2$ 15. $2,200,000 \div 10^8$

16. $3900 \div 10^6$ 17. $3600 \div 10^5$ 18. $17,000 \div 10^3$ 19. $74,000 \div 10^2$ 20. $1000 \div 100$

Multiply the following numbers.

21. 2.5×10^{-3} 22. 3500×10^{-1} 23. 4760×10^{-2} 24. 0.15×10^{-3} 25. 6.5×10^{-4}

26. $150,000 \times 10^{-6}$ 27. $22,500 \times 10^{-3}$ 28. 0.035×10^{-2} 29. $666,000 \times 10^{-7}$ 30. 750×10^{-4}

Divide the following numbers.

31. $\dfrac{100}{10^{-2}}$ 32. $\dfrac{2500}{10^{-1}}$ 33. $\dfrac{4760}{10^{-2}}$ 34. $\dfrac{0.045}{10^{-4}}$ 35. $\dfrac{0.001}{10^{-5}}$

36. $\dfrac{0.05}{10^{-3}}$ 37. $\dfrac{225}{10^{-4}}$ 38. $\dfrac{65.2}{10^{-3}}$ 39. $\dfrac{0.036}{10^{-3}}$ 40. $\dfrac{155,000}{10^{-2}}$

Express the following numbers as a number between 1 and 10 times the proper power of 10. Express answers to three significant figures.

41. 365 42. 6000 43. 15,000 44. 476,000 45. 1,500,000

46. 222 47. 56.7 48. 68.8 49. 75,000,000 50. 1120

51. 0.18 52. 0.00047 53. 0.05 54. 0.0016 55. 0.0000055

56. 0.02 57. 0.0033 58. 0.00056 59. 0.00000000088 60. 0.001

Multiply the following numbers.

61. 300×2000

62. 15×100

63. 100×10^4

64. $50,000 \times 6 \times 10^{-4}$

65. $400 \times 5 \times 10^3 \times 3 \times 10^{-3}$

66. $2.5 \times 10^{-3} \times 10 \times 12 \times 1 \times 10^8$

67. $10^3 \times 10^4 \times 10^{-2}$

68. $1.5 \times 10^3 \times 3.0 \times 10^2$

69. $0.005 \times 10^3 \times 0.0025 \times 10^{-2} \times 10^3$

70. $1000 \times 1000 \times 0.0001 \times 100 \times 10^{-4}$

Divide the following numbers.

71. $\dfrac{10^{12}}{10^3}$

72. $\dfrac{10^5}{10^{-2}}$

73. $\dfrac{15,000}{50}$

74. $\dfrac{6000}{2.5}$

75. $\dfrac{8000 \times 10^2}{40 \times 10^4}$

76. $\dfrac{10^6}{10^8}$

77. $\dfrac{10^9}{10^5}$

78. $\dfrac{250 \times 10^6}{1.25 \times 10^3}$

79. $\dfrac{160,000}{4000}$

80. $\dfrac{100 \times 10^3}{1000 \times 10^{-2}}$

JOB 3-7   Units of Measurement in Electronics

Make the following conversions.

1. 2500 μA to A

2. 2.2 MΩ to kΩ

3. 220 kΩ to MΩ

4. 5.5 kW to W

5. 0.005 V to mV

6. 250 μA to mA

7. 80 mA to μA

8. 2,000,000 V to GV

9. 150 GV to MV

10. 100 nF to μF

11. 20,000 pF to nF

12. 40 nF to μF

13. 0.1 A to mA

14. 4.5 mA to A

15. 65,100 W to kW

16. 3,500,000 Hz to MHz

17. 65 nA to pA

18. 0.0001 μF to nF

19. 0.05 kW to W

20. 0.0003 kW to W

21. 0.004 A to μA

22. 25,000 μA to A

23. 0.00055 V to mV

24. 0.0006 μF to pF

25. 0.036 mV to μV

JOB 3-8   Using Electronics Units of Measurement in Simple Circuits

Solve the following problems for voltage V.

1. $I = 1$ mA, $R = 1$ kΩ, $V =$ _____ V

2. $I = 12$ mA, $R = 3$ kΩ, $V =$ _____ V

3. $I = 300$ μA, $R = 500$ Ω, $V =$ _____ mV

4. $I = 100$ mA, $R = 1.5$ kΩ, $V =$ _____ V

5. $I = 10$ μA, $R = 2.7$ MΩ, $V =$ _____ V

6. $I = 50$ μA, $R = 220$ kΩ, $V =$ _____ V

7. $I = 0.04$ mA, $R = 50$ kΩ, $V =$ _____ V

8. $I = 250$ mA, $R = 0.3$ kΩ, $V =$ _____ V

9. $I = 400$ μA, $R = 750$ kΩ, $V =$ _____ V

10. $I = 1.5$ mA, $R = 220$ Ω, $V =$ _____ mV

11. Use the formula $R_T = \dfrac{R_1 R_2}{R_1 + R_2}$ to find the total resistance of R_1 and R_2 in parallel if

   $R_1 = 1.2\ k\Omega$ and $R_2 = 1800\ \Omega$.

12. An electric heater draws 8 A from the 240-V power line. The power P in watts can be calculated using the formula $P = V \times I$. How many kilowatts of power does the heater use?

13. In a transistor circuit, a 2.4-kΩ emitter resistor R_E has a current of 2.25 mA. Calculate the voltage V across the emitter resistor.

14. The voltage V across a 15-kΩ resistor R is 7.5 V. If $I = \dfrac{V}{R}$, how many microamperes of

   current are flowing in the resistor?

15. If $R = \dfrac{V}{I}$, calculate the value in kilohms of a resistance that has a voltage V of 12 mV and a

   current I of 800 μA.

16. A 300-Ω coil has a current of 2.5 μA. How much voltage in millivolts is dropped across the coil?

17. If $P = V \times I$, how much power P in microwatts is dissipated by a 200-kΩ resistor whose current I is 5 μA and voltage V is 15 V?

18. The time constant of an RC circuit is calculated as $T = RC$. Find T in milliseconds (ms) if $R = 30$ kΩ and $C = 0.05$ μF.

19. The total resistance R_T of a series circuit is calculated using the formula $R_T = R_1 + R_2 + R_3 + \ldots$, etc. Find R_T if $R_1 = 820$ Ω, $R_2 = 12$ kΩ, $R_3 = 180$ Ω, and $R_4 = 22$ kΩ.

20. The total current I_T in a parallel circuit is calculated using the formula $I_T = I_1 + I_2 + I_3 + \ldots$, etc. Find I_T in milliamperes if $I_1 = 500$ μA, $I_2 = 0.02$ A, $I_3 = 1.5$ mA, and $I_4 = 3000$ μA.

JOBS 3-9 to 3-11   Ohm's Law

Solve the following equations for the value of the unknown letter.

1. $2X = 18$ 2. $39 = 3P$ 3. $4X = 48$ 4. $16I = 320$

5. $45X = 90$ 6. $0.5A = 12$ 7. $10C = 80$ 8. $8X = 40$

9. $120 = 15X$ 10. $9Q = 81$ 11. $44 = 11R$ 12. $50 = 0.4F$

13. $22X = 110$ 14. $180 = 10X$ 15. $\frac{1}{2}B = 20$ 16. $5L = 90$

17. $1000X = 10$ 18. $14X = 84$ 19. $25P = 625$ 20. $3 = 18X$

Use the Ohm's law equations $V = I \times R$, $I = \dfrac{V}{R}$, or $R = \dfrac{V}{I}$ to solve the following word problems.

21. A relay operates at 24 V and draws 60 mA. What is its resistance?

22. What is the resistance of a radio if it draws 400 mA of current from a 12-V battery?

23. A 30-W soldering iron has a resistance of 480 Ω. How much current does it draw from the 120-V power line?

24. How much current flows through a 10-kΩ resistor with a voltage drop of 24 V?

25. What is the voltage drop across a 15-kΩ resistor if its current is 30 μA?

26. A 150-Ω resistor has a voltage drop of 60 V. How much current is flowing in the resistor?

27. How much voltage is dropped across a 4.7-kΩ resistor carrying a current of 20 mA?

28. A stereo receiver draws 300 mA from the 120-V power line. What is its resistance?

29. What is the resistance of a lamp if it draws 1.6 A from a 24-Vdc source?

30. An antenna has a resistance of 50 Ω at the feedpoint. What is the current at the feedpoint if the voltage is 223.6 V?

Series Circuits

| JOBS 4-1 to 4-3 | Ohm's Law in Series Circuits |

In each figure, solve for the unknowns listed.

1. In Fig. 4-1, solve for
   R_T, I_1, I_2, I_T, V_1, and V_2.

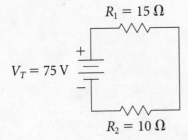

$R_1 = 15\ \Omega$

$V_T = 75\ \text{V}$

$R_2 = 10\ \Omega$

Figure 4-1

3. In Fig. 4-3, solve for
   R_T, I_2, I_T, V_1, V_2, and V_T.

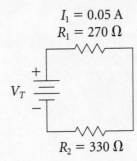

$I_1 = 0.05\ \text{A}$
$R_1 = 270\ \Omega$

V_T

$R_2 = 330\ \Omega$

Figure 4-3

2. In Fig. 4-2, solve for
   R_T, I_1, I_T, V_1, V_2, and V_T.

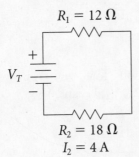

$R_1 = 12\ \Omega$

V_T

$R_2 = 18\ \Omega$
$I_2 = 4\ \text{A}$

Figure 4-2

4. In Fig. 4-4, solve for
   R_T, I_1, I_T, V_1, V_2, and V_T.

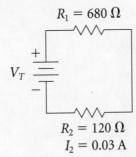

$R_1 = 680\ \Omega$

V_T

$R_2 = 120\ \Omega$
$I_2 = 0.03\ \text{A}$

Figure 4-4

5. In Fig. 4-5, solve for
   R_T, I_1, I_T, and V_T.

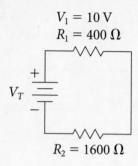

Figure 4-5

8. In Fig. 4-8, solve for
   R_T, I_1, I_3, I_T, V_1, V_2, V_3, and V_T.

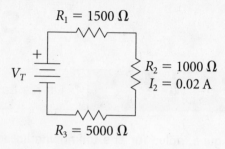

Figure 4-8

6. In Fig. 4-6, solve for
   R_T, I_1, I_2, I_3, I_T, V_1, V_2, and V_3.

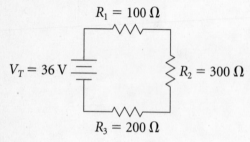

Figure 4-6

9. In Fig. 4-9, solve for
   R_T, I_1, I_2, I_T, V_1, V_2, V_3, and V_T.

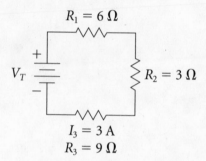

Figure 4-9

7. In Fig. 4-7, solve for
   R_T, I_1, I_2, I_3, I_T, V_1, V_2, and V_3.

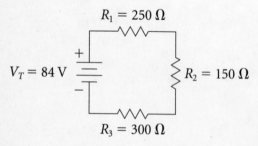

Figure 4-7

10. In Fig. 4-10, solve for
    R_T, I_1, I_2, I_3, I_T, V_2, V_3, and V_T.

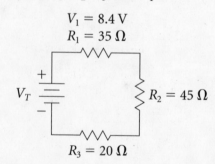

Figure 4-10

11. In Fig. 4-11, solve for
I_2, I_3, I_T, R_1, R_2, R_3, R_T, and V_T.

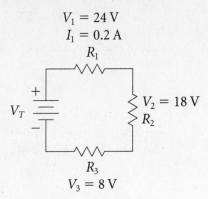

$V_1 = 24$ V
$I_1 = 0.2$ A
R_1

V_T

$V_2 = 18$ V
R_2

R_3
$V_3 = 8$ V

Figure 4-11

14. In Fig. 4-14, solve for
I_1, I_2, I_3, R_1, V_2, R_3, V_T, and R_T.

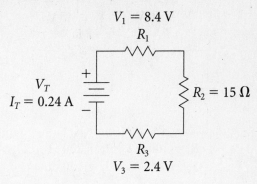

$V_1 = 8.4$ V
R_1

V_T
$I_T = 0.24$ A

$R_2 = 15$ Ω

R_3
$V_3 = 2.4$ V

Figure 4-14

12. In Fig. 4-12, solve for
I_1, I_2, I_3, I_T, R_1, V_3, R_T, and V_T.

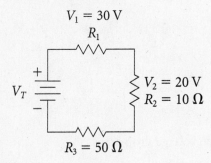

$V_1 = 30$ V
R_1

V_T

$V_2 = 20$ V
$R_2 = 10$ Ω

$R_3 = 50$ Ω

Figure 4-12

15. In Fig. 4-15, solve for
I_1, I_2, I_3, V_1, V_2, R_3, V_T, and R_T.

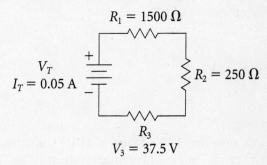

$R_1 = 1500$ Ω

V_T
$I_T = 0.05$ A

$R_2 = 250$ Ω

R_3
$V_3 = 37.5$ V

Figure 4-15

13. In Fig. 4-13, solve for
I_1, I_2, I_T, V_1, R_2, V_3, and V_T.

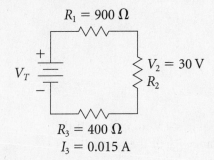

$R_1 = 900$ Ω

V_T

$V_2 = 30$ V
R_2

$R_3 = 400$ Ω
$I_3 = 0.015$ A

Figure 4-13

16. In Fig. 4-16, solve for
R_T, I_1, I_2, I_3, I_4, I_T, V_1, V_2, V_3, and V_4.

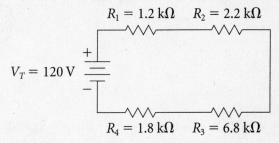

$R_1 = 1.2$ kΩ $R_2 = 2.2$ kΩ

$V_T = 120$ V

$R_4 = 1.8$ kΩ $R_3 = 6.8$ kΩ

Figure 4-16

17. In Fig. 4-17, solve for
    $I_1, I_2, I_4, I_T, R_1, R_2, V_3, V_4, R_T$, and V_T.

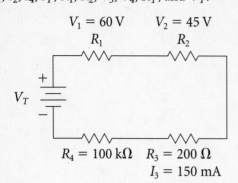

Figure 4-17

18. In Fig. 4-18, solve for
    $I_1, I_2, I_3, I_4, I_5, V_1, R_2, R_3, V_4, V_5, R_T$, and V_T.

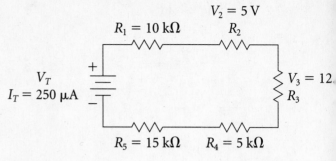

Figure 4-18

Solve the following problems.

19. A 120- and 180-Ω resistor are in series. How much total voltage V_T is required to produce a total current I_T of 150 mA?

20. Three resistors of 10, 15, and 25 Ω are connected in series. If the voltage drop across the 15-Ω resistor is 36 V, how much is the total voltage V_T?

21. A 40-Ω resistor R_1 is in series with an unknown resistor R_2. If the total current I_T equals 200 mA and the voltage drop across R_2 is 16 V, what is the resistance of R_2?

22. A 12-V battery is in series with a resistor R_1 and a radio that requires 9 V to operate. If the radio draws a total current I_T of 75 mA, what is the total resistance R_T of the circuit?

23. Two resistors R_1 and R_2 are used in series with a voltage source V_T for the purpose of biasing the base of a transistor. If $R_1 = 10$ kΩ, $R_2 = 2.2$ kΩ, and $V_2 = 1.8$ V, what are the values of I_T, R_T, V_1, and V_T?

24. A 1.2-kΩ resistor R_1 is in series with an unknown resistor R_2. The total current I_T is 12.5 mA. If the voltage drop across R_2 is 9 V, what are the values of V_1, R_T, and V_T?

25. Three resistors of 1.5, 4.5, and 2 kΩ are in series. If the current through the 2-kΩ resistor is 4.5 mA, what are the values for R_T, I_T, and V_T?

26. A string of 15 identical light bulbs are connected in series. The voltage across each bulb is 8 V and the total current I_T is 320 mA. How much resistance does each light bulb have? What are the values for V_T and R_T?

27. Three identical resistors R_1, R_2, and R_3 in series each have a current of 40 mA. If the total voltage V_T equals 108 V, what is the value of each resistor?

28. What level of current does the Underwriters Laboratories (UL) set for its leakage tests?

29. When dealing with electric shock, what is a representative value of the let-go current?

30. A woman's hand accidentally comes in contact with a 120-V power line. How much current will flow through the woman's body to ground if her body resistance is 5 kΩ and her shoe resistance is 1 kΩ?

JOBS 4-4 to 4-6	Line Currents

Refer to Fig. 4-19. For each of the following, find I_1, I_2, and I_N.

1. $R_1 = 2\ \Omega$ and $R_2 = 3\ \Omega$.

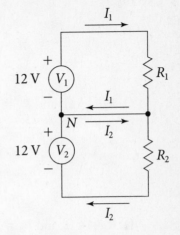

2. $R_1 = 6\ \Omega$ and $R_2 = 8\ \Omega$.

3. $R_1 = 4\ \Omega$ and $R_2 = 4\ \Omega$.

Figure 4-19

4. $R_1 = 3\ \Omega$ and $R_2 = 12\ \Omega$.

5. $R_1 = 1.5\ \Omega$ and $R_2 = 1\ \Omega$.

6. $R_1 = 0.6\ \Omega$ and $R_2 = 0.4\ \Omega$.

7. $R_1 = 0.8\ \Omega$ and $R_2 = 0.8\ \Omega$.

8. $R_1 = 4\ \Omega$ and $R_2 = 6\ \Omega$.

Refer to Fig. 4-20. For each of the following find I_1, I_2, and I_N.

9. $R_1 = 2\ \Omega$, $R_2 = 1\ \Omega$, and $R_3 = 4\ \Omega$.

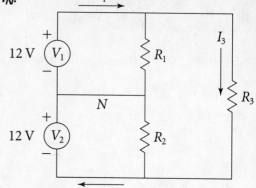

Figure 4-20

10. $R_1 = 4\ \Omega$, $R_2 = 6\ \Omega$, and $R_3 = 2\ \Omega$.

11. $R_1 = 3\ \Omega$, $R_2 = 4\ \Omega$, and $R_3 = 6\ \Omega$.

12. $R_1 = 1\ \Omega$, $R_2 = 1\ \Omega$, and $R_3 = 2\ \Omega$.

13. $R_1 = 2\ \Omega$, $R_2 = 0.5\ \Omega$, and $R_3 = 12\ \Omega$.

14. $R_1 = 3\ \Omega$, $R_2 = 3\ \Omega$, and $R_3 = 3\ \Omega$.

15. $R_1 = 8\ \Omega$, $R_2 = 6\ \Omega$, and $R_3 = 4\ \Omega$.

16. What is a GFCI device? How does it protect personnel against electric shock?

17. Three resistors R_1, R_2, and R_3 are in series. Resistor $R_1 = 100\ \Omega$, $R_2 = 60\ \Omega$, and $R_3 = 240\ \Omega$. If $V_3 = 30\ V$, find I_T, V_1, V_2, R_T, and V_T.

18. A 2-kΩ resistor, R_1 is in series with an 8-kΩ resistor R_2. If the total current I_T is 1.6 mA, find V_1, V_2, V_T, and R_T.

19. A 15-Ω resistor R_1 is in series with a 12-V lamp that needs 200 mA of current to light with full brilliance. How much total voltage V_T must be applied to the circuit to light the lamp with full brilliance?

20. A 120-Vdc source supplies a current of 50 mA to three equal resistances R_1, R_2, and R_3 in series. What is the resistance of each resistor?

JOBS 4-7 and 4-8   Formulas Involving Addition and Subtraction

Solve each of the following equations for the value of the unknown letter.

1. $X + 8 = 20$

2. $P + 4 = 19$

3. $24 + X = 41$

4. $2 + X = 8$

5. $113 + Q = 204$

6. $36 + L = 48$

7. $Z + 62 = 100$

8. $Z + 31 = 33$

9. $M + 68 = 122$

10. $F + 13 = 16$

11. $36 + A + 22 = 100$

12. $112 + B + 14 = 212$

13. $88 + Q + 3 = 101$

14. $3\frac{1}{3} + X + 4\frac{2}{3} = 30$

15. $L + 3 = 50$

16. $26 + d = 500$

17. $I + \dfrac{3}{4} = 6$

18. $P + 6\dfrac{3}{8} = 10\dfrac{1}{4}$

19. $C + 31.6 = 47.8$

20. $A + 2.4 + 8\dfrac{3}{5} = 30$

21. $X - 15 = 41$

22. $V - 21 = 66$

23. $A - 13 - 19 = 105$

24. $X - 0.45 = 13.65$

25. $R - 12 = 133$

26. $Q - 133 + 69 = 24$

27. $G - 6 + 15 = 30$

28. $H + 60 - 22 = 100$

29. $I - 2\frac{1}{2} = 30\frac{3}{4}$

30. $X - 25 = 25$

31. $B - 10 + 6 - 4 = 33$

32. $X - 6 = 12$

33. $B + 8 - 30 = 111$

34. $Z - 60 = 30$

35. $Y + 6 - 3 + 2 - 4 = 20$

36. $L - 3 = 4$

37. $I - 30 = 0$

38. $C - 20 + 20 - 4 = 1$

39. $X - 7 = 13$

40. $A - 21 + 29 - 3 = 104$

Solve each of the following.

41. Using the formula $V_T = V_1 + V_2$, find V_2 if $V_T = 12$ V and $V_1 = 6.7$ V.

42. Using the formula $I_T = I_1 + I_2 + I_3$, solve for I_1 if $I_T = 12$ A, $I_2 = 3$ A, and $I_3 = 4.5$ A.

43. Using the formula $P_T = P_1 + P_2 + P_3$, solve for P_3 if $P_T = 300$ mW, $P_1 = 100$ mW, and $P_2 = 75$ mW.

44. Using the formula $C_T = C_1 + C_2 + C_3$, solve for C_2 if $C_T = 0.001$ F, $C_1 = 0.0001$ F, and $C_3 = 0.0002$ F.

45. Using the formula $L_T = L_1 + L_2 + 2L_M$, solve for L_M if $L_T = 110$ mH, $L_1 = 30$ mH, and $L_2 = 60$ mH.

46. Using the formula $L_T = L_1 + L_2 - 2L_M$, solve for L_M if $L_T = 500$ μH, $L_1 = 300$ μH, and $L_2 = 450$ μH.

47. Using the formula $R_X = \dfrac{V_F}{I_F} - r_m$, solve for r_m if $R_X = 198$ kΩ, $V_F = 10$ V, and

$I_F = 50$ μA.

48. Using the formula $I_S = I_T - I_M$, solve for I_M if $I_S = 9.95$ mA and $I_T = 10$ mA.

49. Using the formula $R_T = R_1 + R_2 + R_3$, find R_2 if $R_T = 5.5$ kΩ, $R_1 = 800$ Ω, and $R_3 = 2.3$ kΩ.

50. Using the formula $C_2 = C_T - C_1 - C_3$, find C_1 if $C_T = 0.003$ μF, $C_2 = 0.001$ μF, and $C_3 = 0.0005$ μF.

JOBS 4-9 and 4-10   Series Circuits

In each figure, solve for the unknowns listed.

1. In Fig. 4-21, solve for
   I_T, R_T, R_1, and V_1.

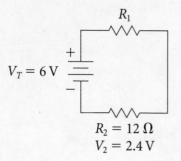

Figure 4-21

3. In Fig. 4-23, solve for
   V_2, I_T, R_T, and R_1.

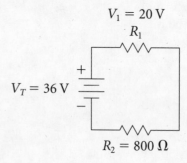

Figure 4-23

2. In Fig. 4-22, solve for
   I_T, R_T, R_2, and V_2.

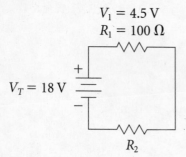

Figure 4-22

4. In Fig. 4-24, solve for
   R_1, V_1, V_2, and V_T.

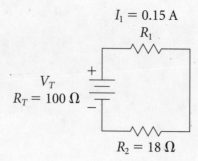

Figure 4-24

5. In Fig. 4-25, solve for
I_T, V_T, V_2, V_3, and R_2.

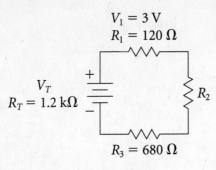

$V_1 = 3$ V
$R_1 = 120$ Ω

V_T
$R_T = 1.2$ kΩ

R_2

$R_3 = 680$ Ω

Figure 4-25

6. In Fig. 4-26, solve for
R_1, R_3, R_T, V_2, and V_T.

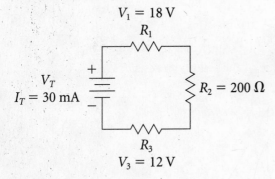

$V_1 = 18$ V
R_1

V_T
$I_T = 30$ mA

$R_2 = 200$ Ω

R_3
$V_3 = 12$ V

Figure 4-26

7. In Fig. 4-27, solve for
R_T, R_3, V_1, V_2, and V_3.

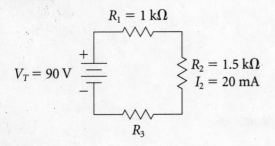

$R_1 = 1$ kΩ

$V_T = 90$ V

$R_2 = 1.5$ kΩ
$I_2 = 20$ mA

R_3

Figure 4-27

8. In Fig. 4-28, solve for
V_1, I_T, R_T, R_2, and R_3.

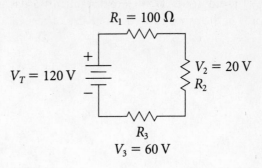

$R_1 = 100$ Ω

$V_T = 120$ V

$V_2 = 20$ V
R_2

R_3
$V_3 = 60$ V

Figure 4-28

9. In Fig. 4-29, solve for
R_2, I_T, V_T, V_1, V_3, and V_4.

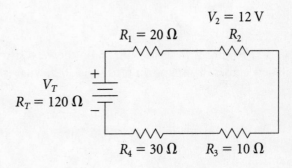

$V_2 = 12$ V
$R_1 = 20$ Ω R_2

V_T
$R_T = 120$ Ω

$R_4 = 30$ Ω $R_3 = 10$ Ω

Figure 4-29

10. In Fig. 4-30, solve for
R_T, R_4, V_1, V_2, V_3, R_4, and V_4.

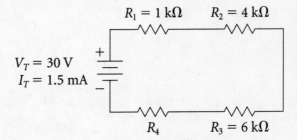

$R_1 = 1$ kΩ $R_2 = 4$ kΩ

$V_T = 30$ V
$I_T = 1.5$ mA

R_4 $R_3 = 6$ kΩ

Figure 4-30

Solve each of the following problems.

11. A 6-V radio needs 200 mA to operate. In order to operate the radio from a 15-V power supply at the same current, what size of resistance must be connected in series with it?

12. How much resistance must be connected in series with a 150-Ω resistor to limit the current from a 12-V battery to 30 mA?

13. A 30-Ω resistor is in series with an unknown resistor. If the total voltage V_T is 120 V and the voltage across the unknown resistor is 30 V, what is the value of the unknown resistor?

14. A 12-V relay requires 50 mA to operate. How much resistance must be added in series with the relay if it is to be operated from a 30 Vdc source?

15. A motor drawing 30 A is connected to the 240-V power line through wires that have a total resistance R_T of 0.2 Ω. How much voltage is available at the motor?

Parallel Circuits

| JOBS 5–1 and 5–2 | V, I, and R in Parallel Circuits |

In each figure, solve for the unknowns listed.

1. In Fig. 5-1, solve for I_T, V_1, V_2, and R_T.

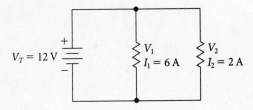

$V_T = 12\ V$ V_1 $I_1 = 6\ A$ V_2 $I_2 = 2\ A$

Figure 5-1

3. In Fig. 5-3, solve for I_T, V_T, and R_T.

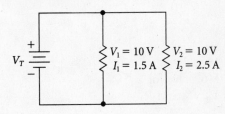

V_T $V_1 = 10\ V$ $I_1 = 1.5\ A$ $V_2 = 10\ V$ $I_2 = 2.5\ A$

Figure 5-3

2. In Fig. 5-2, solve for I_T, V_1, V_2, and R_T.

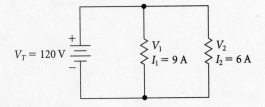

$V_T = 120\ V$ V_1 $I_1 = 9\ A$ V_2 $I_2 = 6\ A$

Figure 5-2

4. In Fig. 5-4, solve for I_T, V_1, V_T, and R_T.

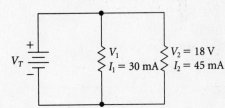

V_T V_1 $I_1 = 30\ mA$ $V_2 = 18\ V$ $I_2 = 45\ mA$

Figure 5-4

5. In Fig. 5-5, solve for
I_T, V_T, V_1, V_3, and R_T.

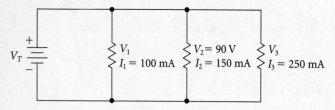

Figure 5-5

6. In Fig. 5-6, solve for
I_T, V_1, V_2, V_3, and R_T.

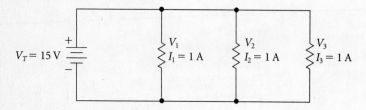

Figure 5-6

7. In Fig. 5-7, solve for
V_T, V_1, V_2, V_3, I_2, I_T, R_3, and R_T.

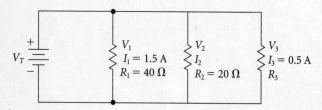

Figure 5-7

8. In Fig. 5-8, solve for
   V_T, V_1, V_3, I_1, I_2, R_3, I_T, and R_T.

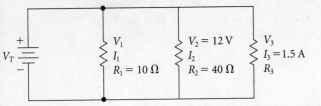

Figure 5-8

9. In Fig. 5-9, solve for
   V_T, V_2, V_3, I_1, I_2, I_T, R_3, and R_T.

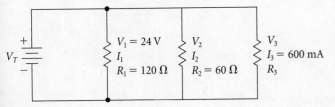

Figure 5-9

10. In Fig. 5-10, solve for
    V_T, V_1, V_2, V_3, R_1, I_2, I_T, and R_T.

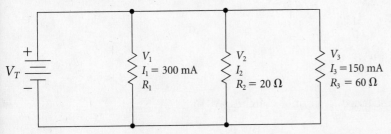

Figure 5-10

In Fig. 5-11 the switches SW1–SW5 are used to connect a variety of different kitchen appliances to the 120-V power line. The current value shown for each appliance exists only if the switch controlling that appliance is closed. If a switch is open that appliance does not draw any current. If the total current I_T exceeds 25 A, which is the current rating of the fuse, the fuse will blow and none of the appliances can be operated. In reference to Fig. 5-11, answer Questions 11 to 15.

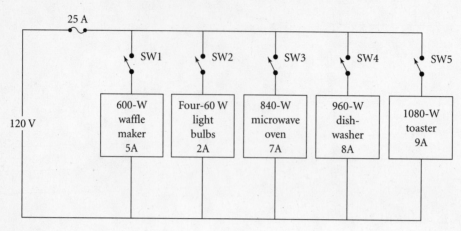

Figure 5-11

11. Calculate the resistance of each individual appliance in Fig. 5-11.

12. Calculate the total current I_T and total resistance R_T if only switches SW2, SW3, and SW4 are closed.

13. Calculate the total current I_T and total resistance R_T if only switches SW1 and SW5 are closed.

14. Will the fuse blow if all switches are closed?

15. Will the fuse blow if switches SW1, SW2, SW4, and SW5 are closed?

Solve each of the following.

16. Six 75-W light bulbs are connected in parallel with a 120-V power line. If each bulb draws a current of 625 mA, how much is (*a*) the total current I_T, (*b*) the voltage across each bulb, and (*c*) the total resistance R_T of the circuit.

17. An electric heater drawing 12.5 A, a 60-W light bulb drawing 0.5 A and a radio drawing 100 mA are connected in parallel with the 120-V power line. (*a*) How much voltage is across each device? (*b*) What is the total current I_T drawn by all three devices? (*c*) What is the total resistance R_T of the circuit?

18. A 10-Ω resistor carrying a current of 3.6 A is in parallel with a 15-Ω resistor. Calculate V_T, I_T, and R_T.

19. A 120-V line is protected with a 30-A fuse. How many 150-W light bulbs can be connected in parallel with the 120-V line without blowing the fuse? Each 150-W bulb draws 1.25 A of current.

20. Three resistors R_1, R_2, and R_3 are in parallel with a voltage source V_T. $R_1 = 150\ \Omega$, $R_2 = 300\ \Omega$, and $R_3 = 100\ \Omega$. If $I_2 = 50$ mA, find V_T, V_1, V_3, I_1, I_3, I_T, and R_T.

21. Explain the differences between a series and a parallel circuit.

22. Refer to Fig. 5-11. Find R_T with (a) only switch SW1 closed, (b) switches SW1 and SW2 closed, and (c) switches SW1, SW2, and SW3 closed. Explain why R_T decreases as more branches are connected across the 120-V line.

| JOBS 5-3 and 5-5 | Addition and Subtraction of Fractions |

Add the following fractions and express the answers in lowest terms.

1. $\dfrac{1}{3} + \dfrac{1}{4}$

2. $\dfrac{2}{9} + \dfrac{2}{3}$

3. $\dfrac{1}{4} + \dfrac{1}{2}$

4. $\dfrac{2}{5} + \dfrac{1}{3}$

5. $\dfrac{1}{2} + \dfrac{3}{8}$

6. $\dfrac{1}{6} + \dfrac{2}{3} + \dfrac{1}{8}$

7. $\dfrac{3}{10} + \dfrac{3}{20} + \dfrac{1}{5}$

8. $\dfrac{2}{6} + \dfrac{3}{9} + \dfrac{1}{8}$

9. $\dfrac{3}{16} + \dfrac{1}{8} + \dfrac{5}{8}$

10. $\dfrac{1}{11} + \dfrac{2}{8} + \dfrac{4}{9}$

11. $\dfrac{45}{100} + \dfrac{7}{50} + \dfrac{16}{25}$

12. $\dfrac{3}{12} + \dfrac{1}{3} + \dfrac{2}{9} + \dfrac{1}{4}$

13. $\dfrac{2}{7} + \dfrac{3}{7} + 1$

14. $3\dfrac{3}{8} + 2\dfrac{1}{4} + 4\dfrac{1}{2}$

15. $1\dfrac{3}{4} + 2\dfrac{1}{3} + 3\dfrac{1}{9}$

Subtract the following fractions and express the answers in lowest terms.

16. $\dfrac{7}{8} - \dfrac{1}{4}$

17. $\dfrac{13}{16} - \dfrac{1}{2}$

18. $\dfrac{3}{7} - \dfrac{1}{7}$

19. $\dfrac{15}{20} - \dfrac{3}{4}$

20. $\dfrac{4}{9} - \dfrac{1}{3}$

21. $\dfrac{6}{13} - \dfrac{2}{39}$ 22. $\dfrac{4}{10} - \dfrac{1}{4}$ 23. $\dfrac{6}{7} - \dfrac{2}{3}$ 24. $\dfrac{8}{11} - \dfrac{3}{7}$ 25. $\dfrac{25}{75} - \dfrac{3}{25}$

26. $\dfrac{2}{3} - \dfrac{2}{6}$ 27. $5\dfrac{6}{8} - 2\dfrac{1}{3}$ 28. $14\dfrac{1}{9} - 2\dfrac{3}{4}$ 29. $6\dfrac{1}{3} - 4\dfrac{2}{4}$ 30. $7\dfrac{2}{7} - 5\dfrac{1}{8}$

Solve each of the following.

31. The branch currents in a parallel circuit are $1\frac{2}{3}$ A, $3\frac{1}{2}$ A, and $2\frac{1}{4}$ A. How much is the total current I_T?

32. Two resistors in parallel have a total current I_T of $2\frac{1}{2}$ A. If one branch has a current of $1\frac{3}{8}$ A, how much current is left to flow in the other branch?

33. Three resistors in series have voltage drops of $6\frac{3}{8}$ V, $2\frac{1}{3}$ V, and $5\frac{1}{2}$ V. How much is the total voltage V_T?

34. The inside diameter of an electric conduit is $1\frac{7}{32}$ in. If the thickness of the conduit is $\frac{1}{8}$ in, what is the outside diameter of the conduit?

35. A $6\frac{2}{9}$-in length of No. 18 gage solid hookup wire has had $1\frac{5}{16}$ of an inch of insulation removed from one end. What is the length of wire that still retains the insulation?

36. What is the difference in length between a $\frac{3}{8}$-in-long machine screw and a $\frac{5}{16}$-in-long machine screw?

37. The bottom $3\frac{1}{8}$ ft of a 10-ft tower section is below ground level. What length of tower is above ground level?

38. A parallel circuit is protected by a $\frac{5}{8}$-A fuse. Will the fuse carry two branch currents of $\frac{3}{16}$ A and $\frac{1}{2}$ A without blowing?

39. Two resistors in series receive a total voltage V_T of 12 V. If one resistor has a voltage of $3\frac{5}{16}$ V, how much voltage does the other resistor have?

40. A parallel circuit has the following branch currents: $I_1 = \frac{2}{3}$ A, $I_2 = \frac{1}{2}$ A, $I_3 = 1\frac{3}{8}$ A, and $I_4 = 1\frac{1}{4}$ A. How much is the total current I_T?

JOBS 5–6 and 5–7   Solution of Equations Involving Fractions

Find the value of the unknown letter in each problem.

1. $\dfrac{X}{2} = 18$

2. $\dfrac{a}{1.5} = 30$

3. $\dfrac{Q}{2} = 6$

4. $\dfrac{X}{4} = 28$

5. $\dfrac{L}{3} = 54$

6. $\dfrac{X}{6} = 72$

7. $\dfrac{B}{2.5} = 50$

8. $\dfrac{C}{60} = 0.1$

9. $7 = \dfrac{C}{3}$

10. $100 = \dfrac{X_L}{5}$

Solve each equation for the value of the unknown letter.

11. $\dfrac{X}{3} = \dfrac{6}{8}$

12. $\dfrac{X}{9} = \dfrac{2}{4}$

13. $\dfrac{3}{4} = \dfrac{X}{6}$

14. $\dfrac{10}{b} = \dfrac{15}{20}$

15. $\dfrac{1}{5} = \dfrac{3}{X}$

16. $\dfrac{14}{20} = \dfrac{6}{a}$

17. $\dfrac{4X}{8} = \dfrac{3}{12}$

18. $\dfrac{3}{9} = \dfrac{2B}{6}$

19. $\dfrac{4}{2X} = \dfrac{3}{13}$

20. $\dfrac{30}{150} = \dfrac{6}{3C}$

21. $\dfrac{45}{90} = \dfrac{2P}{60}$

22. $\dfrac{4}{3X} = \dfrac{1}{10}$

23. $\dfrac{11}{2} = \dfrac{a}{4}$

24. $\dfrac{20}{3X} = \dfrac{1}{9}$

25. $\dfrac{10C}{5} = \dfrac{2}{20}$

26. $\dfrac{24}{X} = 3$

27. $\dfrac{4}{R} = \dfrac{0.5}{3}$

28. $\dfrac{6}{R} = 40$

29. $\dfrac{8}{V} = \dfrac{20}{32}$

30. $3 = \dfrac{12}{0.6P}$

Solve each of the following.

31. In the formula $I = \dfrac{P}{V}$, find V if $P = 150$ and $I = 1.25$.

32. In the formula $Q = \dfrac{X_L}{R}$, find X_L if $Q = 60$ and $R = 2 \, \Omega$.

33. In the formula $\cos \theta = \dfrac{R}{Z}$, find Z if $\cos \theta = 0.707$ and $R = 50$.

34. In a transformer the following relationship is true: $\dfrac{I_S}{I_P} = \dfrac{V_P}{V_S}$. Find V_S if

    $I_S = 10$ A, $I_P = 0.5$ A, and $V_P = 120$ V.

35. Using the formula from Prob. 34, find I_P if $V_P = 120$, $V_S = 720$, and $I_S = 0.1$

36. In a series voltage divider, the following relationship is true: $\dfrac{R_1}{R_T} = \dfrac{V_1}{V_T}$. Find V_1 if

    $R_1 = 1 \, k\Omega$, $R_T = 4 \, k\Omega$, and $V_T = 36$ V.

37. Using the formula from Prob. 36, find R_T if $R_1 = 200$, $V_1 = 6$, and $V_T = 24$.

38. In a parallel circuit the following relationship is true: $\dfrac{I_1}{I_T} = \dfrac{R_T}{R_1}$. Find I_1 if $R_1 = 10\ \Omega$,

$R_T = 6\ \Omega$, and $I_T = 12$ A.

39. In a Wheatstone bridge the following relationship is true: $\dfrac{R_X}{R_S} = \dfrac{R_1}{R_2}$. Find R_2 if $R_X = 615\ \Omega$,

$R_S = 6150\ \Omega$, and $R_1 = 1000\ \Omega$.

40. Using the formula from Prob. 39, find R_X if $R_1 = 10\ k\Omega$, $R_2 = 100\ \Omega$, and $R_S = 672\ \Omega$.

JOBS 5–8 to 5–11   Analyzing Parallel Circuits

In each figure, solve for the unknowns listed.

1. In Fig. 5-12, find R_T.

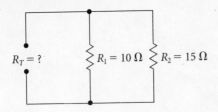

$R_T = ?$ $R_1 = 10\ \Omega$ $R_2 = 15\ \Omega$

Figure 5-12

3. In Fig. 5-14, find R_T.

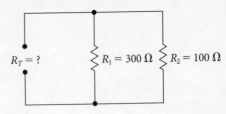

$R_T = ?$ $R_1 = 300\ \Omega$ $R_2 = 100\ \Omega$

Figure 5-14

2. In Fig. 5-13, find R_T.

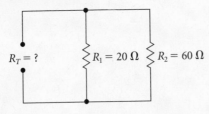

$R_T = ?$ $R_1 = 20\ \Omega$ $R_2 = 60\ \Omega$

Figure 5-13

4. In Fig. 5-15, find R_1.

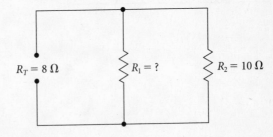

$R_T = 8\ \Omega$ $R_1 = ?$ $R_2 = 10\ \Omega$

Figure 5-15

5. In Fig. 5-16, find R_T.

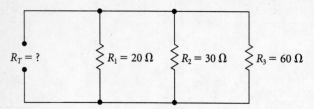

Figure 5-16

6. In Fig. 5-17, find R_T.

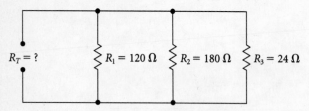

Figure 5-17

7. In Fig. 5-18, find R_T.

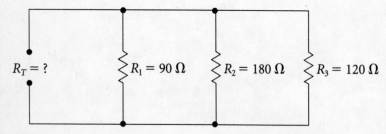

Figure 5-18

8. In Fig. 5-19, find R_3.

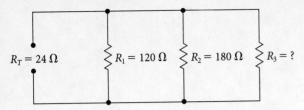

Figure 5-19

9. In Fig. 5-20, find R_T.

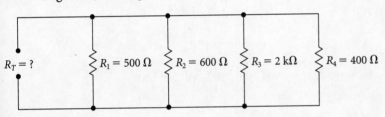

Figure 5-20

10. In Fig. 5-21, find R_T.

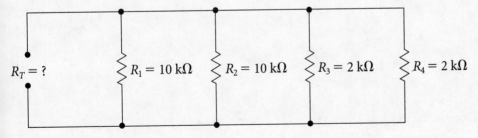

Figure 5-21

Solve each of the following.

11. In Fig. 5-12, what value of total voltage V_T will produce a total current I_T of 5 A?

12. In Fig. 5-13, what value of total voltage V_T will produce a total current I_T of 800 mA?

13. In Fig. 5-14, what value of total voltage V_T will produce a total current I_T of 240 mA?

14. In Fig. 5-16, what value of total voltage V_T will produce a total current I_T of 1.6 A?

15. In Fig. 5-17, what value of total voltage V_T will produce a total current I_T of 2 A?

16. What value of resistance must be connected in parallel with a 2.2-kΩ resistor to obtain a total resistance R_T of 687½ Ω?

17. How many 2.7-kΩ resistors must be connected in parallel to obtain a total resistance R_T of 180 Ω?

18. A 1-kΩ resistor R_1 is in parallel with an unknown resistor R_2. Find the value of R_2 if $R_T = 600 \ \Omega$ and $I_1 = 24$ mA.

19. A 300-Ω resistor R_1 and a 150-Ω resistor R_2 are in parallel with an unknown resistor R_3. Find the value of R_3 if $I_T = 400$ mA and $I_1 = 106.67$ mA.

20. Five equal resistor values in parallel have a total resistance R_T of 36 Ω. Find R_T if two resistors are removed from the circuit.

21. In Fig. 5-22, find
    the values for I_1, I_2, R_T, and V_T.

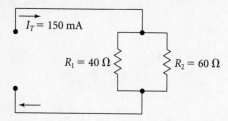

Figure 5-22

22. In Fig. 5-23, find
    the values for I_1, I_2, R_T, and V_T.

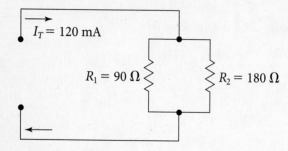

Figure 5-23

23. In Fig. 5-24, find
    the values for I_1, I_2, R_T, and V_T.

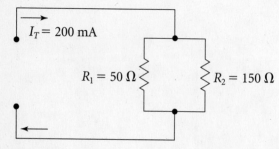

Figure 5-24

78

24. In Fig. 5-25, find
    the values for I_1, I_2, I_3, R_T, and V_T.

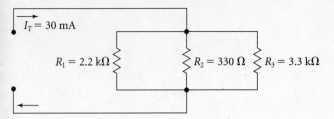

Figure 5-25

25. In Fig. 5-26, find
    the values for I_1, I_2, I_3, R_T, and V_T.

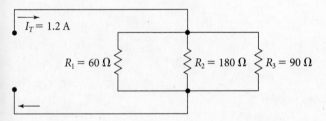

Figure 5-26

26. In Fig. 5-27, find
    R_T, V_T, I_1, I_2, and I_3.

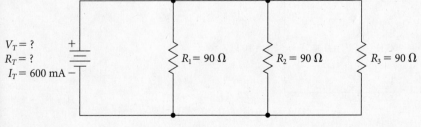

Figure 5-27

27. In Fig. 5-28, find
    R_2, I_T, I_1, I_2, and I_3.

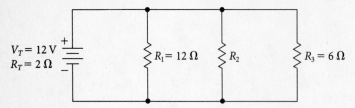

$V_T = 12\,\text{V}$
$R_T = 2\,\Omega$

$R_1 = 12\,\Omega$ R_2 $R_3 = 6\,\Omega$

Figure 5-28

28. In Fig. 5-29, find
    V_T, I_T, I_1, R_1, and I_3.

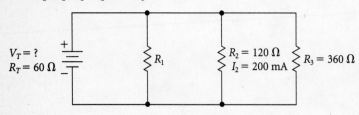

$V_T = ?$
$R_T = 60\,\Omega$

R_1 $R_2 = 120\,\Omega$ $R_3 = 360\,\Omega$
       $I_2 = 200\,\text{mA}$

Figure 5-29

29. In Fig. 5-30, find
    R_T, V_T, I_1, I_2, I_3, and I_4.

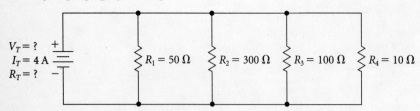

$V_T = ?$
$I_T = 4\,\text{A}$
$R_T = ?$

$R_1 = 50\,\Omega$ $R_2 = 300\,\Omega$ $R_3 = 100\,\Omega$ $R_4 = 10\,\Omega$

Figure 5-30

30. In Fig. 5-31, find
R_T, I_T, I_1, I_2, I_3, and I_4.

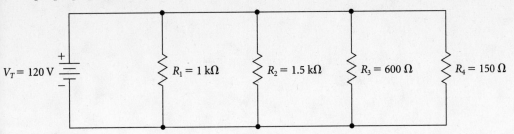

Figure 5-31

Introduction to Electricity

6
c h a p t e r

Solving Combination Circuits

Solve each of the following.

1. In Fig. 6-1, solve for
   R_T, I_T, I_1, I_2, I_3, V_1, V_2, and V_3.

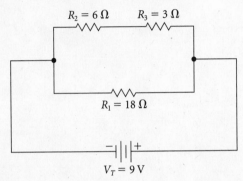

Figure 6-1

3. In Fig. 6-3, solve for
   R_T, I_T, I_1, I_2, I_3, V_1, V_2, and V_3.

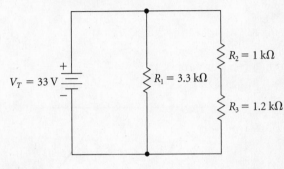

Figure 6-3

2. In Fig. 6-2, solve for
   R_T, I_T, I_1, I_2, I_3, V_1, V_2, and V_3.

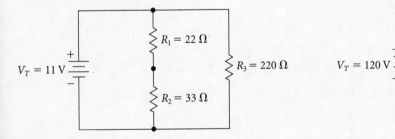

Figure 6-2

4. In Fig. 6-4, solve for
   R_T, I_T, I_1, I_2, I_3, I_4, V_1, V_2, V_3, and V_4.

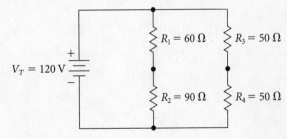

Figure 6-4

5. In Fig. 6-5, solve for
   R_T, I_T, I_1, I_2, I_3, I_4, V_1, V_2, V_3, and V_4.

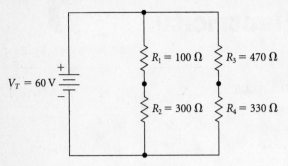

Figure 6-5

6. In Fig. 6-6, solve for
   R_T, I_T, I_1, I_2, I_3, I_4, I_5, V_1, V_2, V_3, V_4, and V_5.

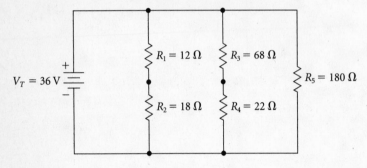

Figure 6-6

7. In Fig. 6-7, solve for
   R_T, I_T, I_1, I_2, I_3, I_4, I_5, I_6, V_1, V_2, V_3, V_4, V_5, and V_6.

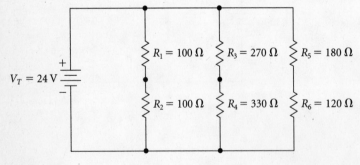

Figure 6-7

8. In Fig. 6-8, solve for
   R_T, I_T, I_1, I_2, I_3, I_4, V_1, V_2, V_3, and V_4.

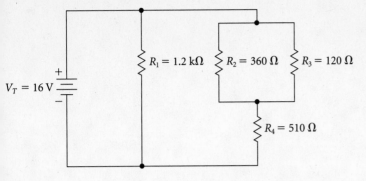

Figure 6-8

9. In reference to Fig. 6-1, how are R_T, I_T, I_1, and I_2 affected if R_1 is removed from the circuit?

10. In reference to Fig. 6-4, how are R_T, I_T, I_1, I_2, I_3, and I_4 affected if another 100-Ω resistor is connected directly across the terminals of the voltage source?

11. In Fig. 6-9, solve for
    R_T, I_T, I_1, I_2, I_3, V_1, V_2, and V_3.

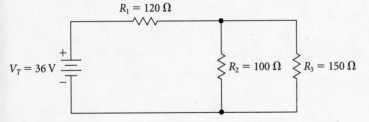

Figure 6-9

12. In Fig. 6-10, solve for
    $R_T, I_T, I_1, I_2, I_3, V_1, V_2,$ and V_3.

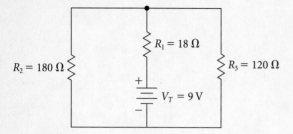

Figure 6-10

13. In Fig. 6-11, solve for
    $R_T, I_T, I_1, I_2, I_3, I_4, V_1, V_2, V_3,$ and V_4.

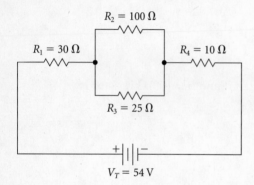

Figure 6-11

14. In Fig. 6-12, solve for
    $R_T, I_T, I_1, I_2, I_3, I_4, V_1, V_2, V_3,$ and V_4.

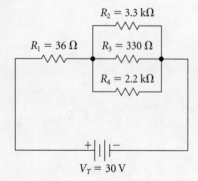

Figure 6-12

15. In Fig. 6-13, solve for
    R_T, I_T, I_1, I_2, I_3, I_4, I_5, V_1, V_2, V_3, V_4, and V_5.

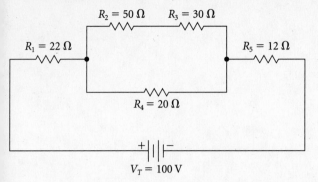

Figure 6-13

16. In Fig. 6-14, solve for
    R_T, I_T, I_1, I_2, I_3, I_4, I_5, I_6, V_1, V_2, V_3, V_4, V_5, and V_6.

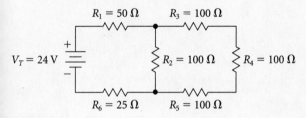

Figure 6-14

17. In Fig. 6-15, solve for
    R_T, I_T, I_1, I_2, I_3, I_4, I_5, V_1, V_2, V_3, V_4, and V_5.

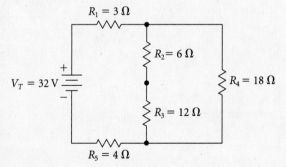

Figure 6-15

18. In Fig. 6-16, solve for
    R_T, I_T, I_1, I_2, I_3, I_4, I_5, V_1, V_2, V_3, V_4, and V_5.

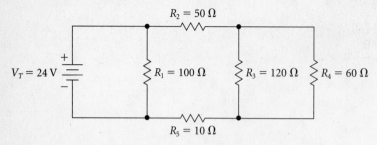

Figure 6-16

19. In reference to Fig. 6-14, how are R_T and V_2 affected if R_4 opens?

20. In reference to Fig. 6-15, how are R_T and V_4 affected if R_3 shorts?

21. A 120-Ω resistor R_1 is in parallel with a 40-Ω resistor R_2. It is desired to have 6 V across these two resistors. If the applied voltage is 24 V, what value of resistance R_1 must be connected in series with R_2 and R_3 in parallel?

22. A 60-Ω resistor R_1 is in series with a 20-Ω resistor R_2. What resistance R_3 must be connected in parallel with R_1 and R_2 to produce a total resistance R_T of 60 Ω?

23. Two light bulbs required 28 V each to operate. Each bulb draws 400 mA. How much resistance R_S must be connected in series with a 120-Vdc source in order to provide 28 V to both bulbs?

24. A 1-kΩ resistor R_1 is in parallel with two equal resistors R_2 and R_3 that are in series. If $R_T = 800$ Ω, what are the values for R_2 and R_3?

25. A 9- and a 3-V lamp are in series across a 12-Vdc source. A 12-V lamp is also connected in parallel with the 12-V source. If the 3-V lamp burns out, which lamps will light?

JOB 6–4   Line Drop

Solve each of the following.

1. In Fig. 6-17, find the line voltage V_{line} and the load voltage V_L. As shown, each line wire has a resistance of 0.25 Ω.

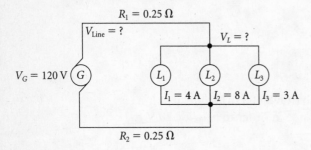

Figure 6-17

2. In Fig. 6-18, find the line voltage V_{line} and the load voltage V_L. Each line wire has a resistance of 0.6 Ω.

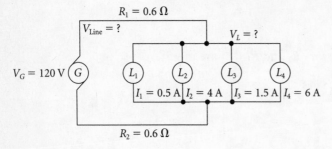

Figure 6-18

3. In Fig. 6-19, find the generator voltage V_G needed to produce a load voltage V_L of 232.5 V. Also, calculate the line voltage V_{line}.

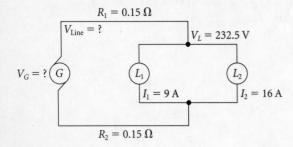

Figure 6-19

4. In Fig. 6-20, find the line voltage V_{line} and the load voltage V_L. Each line wire has a resistance of 0.05 Ω.

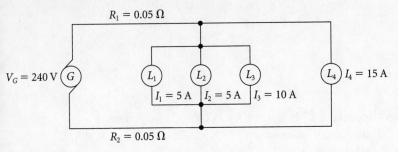

Figure 6-20

5. In Fig. 6-20, what is the load voltage V_L if another 5-A load (L_5) is connected to the circuit?

6. In Fig. 6-20, how much is the line voltage V_{line} if only loads L_1 and L_3 are connected?

7. In Fig. 6-21, find the values of R_1 and R_2.

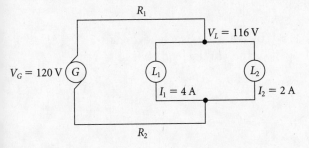

Figure 6-21

8. In Fig. 6-21, what values for R_1 and R_2 will provide a load voltage V_L of 105 V?

9. In Fig. 6-21, another load L_3 is connected in parallel with L_1 and L_2. What value of current would the new load L_3 have to draw to produce a line voltage of 5 V? *Note:* L_1 and L_2 are still connected. (Use the values of R_1 and R_2 calculated in Prob. 7.)

10. In Fig. 6-22, how much current is drawn by load 1 (L_1) if the line voltage is 3.2 V as shown?

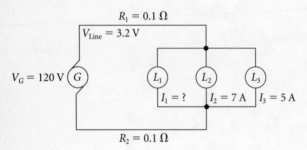

Figure 6-22

11. An 8-A motor is connected to a generator voltage V_G of 115 V by two wires, each having a resistance of 0.35 Ω. How much is (*a*) the line voltage drop V_{line} and (*b*) the load voltage V_L?

12. A 10-A motor is located 250 ft away from a generator whose voltage V_G is 120 V. Calculate the voltage V_L if the wire used to connect the generator to the load has a resistance of 2.6 Ω per 1000 ft.

13. A toaster drawing 9 A and a microwave oven drawing 7 A are connected to the 120-V power line through wires that have a resistance of 0.04 Ω each. How much voltage is available for the toaster and microwave oven? The line is fused at 25 A.

14. In Prob. 13, how much current will flow in the line wires if the toaster shorts? Will the fuse blow?

15. What generator voltage V_G is needed to supply a voltage of 115 V to a skill saw drawing 6 A, a lamp drawing 2.5 A, and a hand drill drawing 1.5 A? The loads are connected to the generator voltage V_G through a 100-ft extension cord whose total line resistance is 0.4 Ω.

JOBS 6–5 and 6–6   Distribution Systems and Ideal Circuit Elements

Solve each of the following.

1. In Fig. 6-23, find the voltage across (*a*) motor M_1, (*b*) motor M_2, and (*c*) motor M_3.

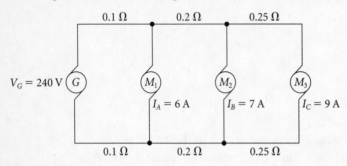

Figure 6-23

2. In Fig. 6-24, find the voltage across (*a*) lamp bank 1 and (*b*) lamp bank 2.

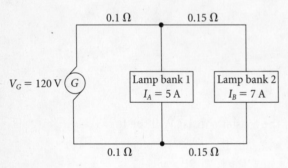

Figure 6-24

3. In Fig. 6-25, find (*a*) the voltage V_B and (*b*) the generator voltage V_G.

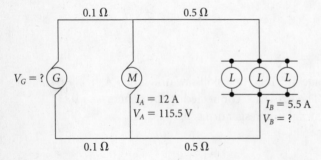

Figure 6-25

4. In Fig. 6-26, find (*a*) the voltage V_A, (*b*) the current I_C, and (*c*) the generator voltage V_G.

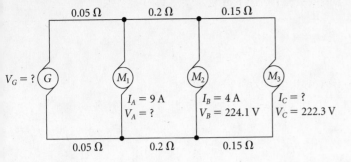

Figure 6-26

5. In Fig. 6-27, find (*a*) the motor voltage V_A and (*b*) the lamp voltage V_B.

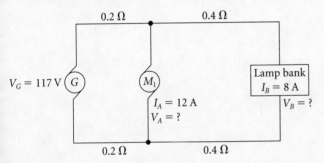

Figure 6-27

6. In Fig. 6-28, each lamp draws 1.2 A of current. Find (*a*) V_A and (*b*) V_B.

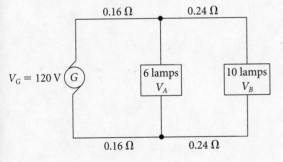

Figure 6-28

Electric Power

7

c h a p t e r

| JOBS 7–1 and 7–2 | Electric Power in Electric Circuits |

Solve each of the following.

1. Calculate the amount of power consumed by an electric heater that draws 10 A from a 120-V power line.

2. A 9-V radio draws 60 mA of current at half volume. How much power is consumed by the radio?

3. A light bulb draws 0.625 A from a 120-V power line. What is the wattage rating of the bulb?

4. A soldering iron draws 250 mA from a 120-V power line. What is the wattage rating of the iron?

5. How much power is consumed by an electric shaver that draws 30 mA from a 120-V power line?

6. A headlight in a car draws 5 A from the battery. If the battery has a voltage of 13.8 V, how much power is consumed by the headlight?

7. A 24-V power supply delivers a current of 0.75 A to a CB radio. How much power is consumed by the CB?

8. Three identical bulbs in a light fixture draw 2.5 A from a 120-V power line. How much power (*a*) is consumed by all three bulbs and (*b*) consumed by each individual bulb?

9. A microwave oven draws 9 A from a 120-V power line when cooking on high power. How much power is being used by the microwave oven?

10. A power amplifier supplies a current of 2.5 A and a voltage of 10 V to a speaker load. How much power is absorbed by the speaker?

11. A carbon resistor carrying a current of 5 mA has a voltage of 50 V. (*a*) How many watts of power will the resistor have to dissipate as heat? (*b*) What is the recommended wattage rating for the resistor in this application?

12. A 100-Ω resistor R_1 and a 150-Ω resistor R_2 are in parallel. If the total current I_T is 60 mA, how much power is dissipated by (a) R_1 and (b) R_2.

13. A clothes dryer draws 16 A from a 240-V power line. Calculate the amount of power used.

14. An amateur radio transmitter delivers 223.6 V and 4.472 A to the feedpoint of its antenna. How much power is being delivered to the antenna?

15. The current in a 200-Ω resistor is 50 mA. Calculate the amount of power dissipated by the resistor if its voltage is 10 V.

16. An 8-A motor, 4-A skill saw, and a 3-A drill are all drawing current from the 120-V power line. How much total power is being used?

17. Two 6-V lamps in series draw 2.5 A of current from a 12-Vdc supply. How much total power is being used by both lamps?

18. Find the total power drawn from a 12-V battery that supplies 6 A of current to the headlights and 1.5 A of current to the dashboard lights.

19. Find the total power used by the following appliances connected to the 120-V power line: A 3-A mixer, an 8-A toaster, a 6.25-A microwave, and a 833.3-mA light bulb.

20. A CD player drawing 0.5 A and an amplifier drawing 4 A are connected to a 13.8-V battery in a car. What is the total power used by the CD player and its amplifier?

21. A 10-Ω resistor R_1 and a 30-Ω resistor R_2 are in series. The total voltage V_T is 12 V. Find (a) the total current I_T, (b) the power used by each resistor, and (c) the total power P_T.

22. A 100-Ω resistor R_1, a 150-Ω resistor R_2, and a 250-Ω resistor R_3 are in series across a total voltage V_T of 20 V. Find (a) the total current I_T, (b) the power used by each resistor, and (c) the total power, P_T.

23. A 12-Ω resistor R_1 and an 18-Ω resistor R_2 are in parallel with a 36-Vdc source. Find (a) the power used by reach resistor and (b) the total power P_T.

24. In Fig. 7-1, find (*a*) the total current I_T, (*b*) the power used by each resistor, and (*c*) the total power P_T.

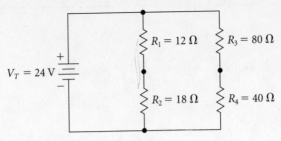

Figure 7-1

25. In Fig. 7-2, find (*a*) the total current I_T, (*b*) the power used by each resistor, and (*c*) the total power P_T.

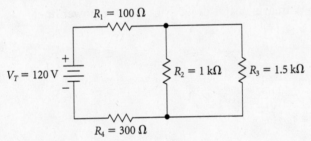

Figure 7-2

JOBS 7–3 and 7–4   Power Formula for Current or Voltage

Answer each of the following.

1. How much current does a 300-W light bulb draw from a 120-V power line?

2. How much current does a 1050-W toaster draw from a 120-V power line?

3. How much current does a 1500-W quartz heater draw from a 120-V power line?

4. A resistor carrying a current of 5 mA dissipates ¼ W of power. How much voltage is across the resistor?

5. A radio transmitter delivers 400 W of power to the feedpoint of an antenna. If the current at the feedpoint is 10 A, how much voltage is at the feedpoint?

6. How much current does a 10-kW electric dryer draw from a 240-V power line?

7. How much current does a 60-W headlamp draw from a 12-V battery?

8. A resistor carrying a current of 40 mA dissipates a power of 432 mW. How much voltage is across the resistor?

9. Six 75-W lights are connected to a 120-V power line. Find (*a*) the current in each bulb and (*b*) the total current I_T.

10. A 250-W television set, a 50-W VCR, and a 100-W light bulb are connected to a 120-V power line. Find (*a*) the current drawn by the television, (*b*) the current drawn by the VCR, (*c*) the current drawn by the light bulb, and (*d*) the total current I_T.

11. A 600-W microwave oven, a 1200-W toaster, a 240-W mixer, and a 150-W refrigerator are connected to a 120-V power line. The line wires providing power to these appliances are protected by a 15-A fuse. Will the fuse blow?

12. Calculate the power dissipated by a 50-Ω resistor whose voltage is 8 V.

13. Calculate the power dissipated by a 1.5-kΩ resistor whose voltage is 3 V.

14. The current carried by a 10-kΩ resistor is 2 mA. How much power is the resistor dissipating as heat?

15. Calculate the power rating of a soldering iron whose resistance is 576 Ω. The soldering iron is operated from a 120-V line.

16. In Fig. 7-3, solve for R_T, I_T, V_1, V_2, P_1, P_2, and P_T.

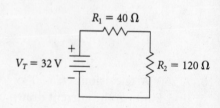

$$R_1 = 40 \, \Omega$$
$$V_T = 32 \, V$$
$$R_2 = 120 \, \Omega$$

Figure 7-3

17. In Fig. 7-4, solve for I_1, I_2, I_T, R_T, P_1, P_2, and P_T.

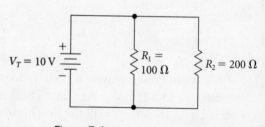

$$V_T = 10 \, V$$
$$R_1 = 100 \, \Omega$$
$$R_2 = 200 \, \Omega$$

Figure 7-4

18. In Fig. 7-5, solve for I_T, V_T, R_2, V_1, V_3, V_4, P_1 P_3, P_4, and P_T.

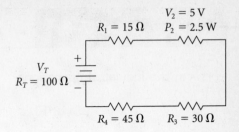

Figure 7-5

19. In Fig. 7-6, solve for I_T, R_T, R_3, I_1, I_2, I_3, P_1, P_2, and P_3.

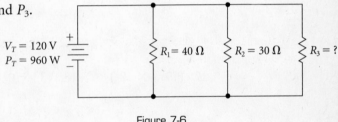

Figure 7-6

20. In Fig. 7-7, solve for R_T, I_T, P_1, P_2, P_3, P_4, P_5, P_6, and P_T.

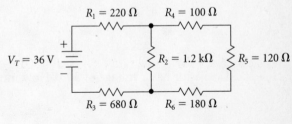

Figure 7-7

JOBS 7-5 and 7-6   Powers of 10

Perform the indicated operations.

1. $(10^2)^2$

2. $(10^4)^2$

3. $(10^1)^3$

4. $(10^5)^3$

5. $(10^{-3})^2$

6. $(10^{-5})^4$

7. $(10^{-3})^{-3}$

8. $(10^6)^{-4}$

9. $(2 \times 10^3)^2$

10. $(1.6 \times 10^2)^2$

11. $(7.5 \times 10^{-3})^2$

12. $(0.003)^2$

13. $(0.016)^2$

14. $(4.0 \times 10^4)^2$

15. $(0.0001)^2$

16. $(2000)^2$

17. $(10^{-3} \times 10^3)^2$

18. $(3 \times 10^4)^2$

19. $(9.0 \times 10^4)^2$

20. $(0.00006)^2$

Solve each of the following.

21. A 100-Ω resistor is carrying a current of 20 mA. How much power is dissipated by the resistor?

22. A 10-kΩ resistor has a voltage of 30 V. How much power is dissipated by the resistor?

23. An 8-Ω load has a voltage of 40 V. How much power is dissipated by the 8-Ω load?

24. A 200-ft length of No. 14 gage copper wire has a resistance of 0.5 Ω. Calculate the power loss in the wire if it carries a current of (*a*) 10 A and (*b*) 20 A.

25. How much power is dissipated by a 144-Ω load that carries a current of 833.3 mA?

26. A 100-Ω resistor in a transistor circuit has a voltage of 10 V. Calculate (*a*) the power dissipated by the resistor and (*b*) the recommended wattage rating of the resistor.

27. The line wires in a power distribution system have a total resistance of 0.1 Ω. If the line voltage V_{Line} is 6 V, how much power is dissipated by the line wires?

28. A 48-Ω light bulb has a current of 2.5 A. What is the power rating of the light bulb?

29. A 50-Ω antenna has a current of 1.5 A. How much power is delivered to the antenna?

30. How much power is dissipated by a 47-Ω resistor whose voltage is 9.4 V?

31. A 50-Ω resistor R_1 and a 30-Ω resistor R_2 are in series. If the series current is 300 mA, find (a) P_1 and P_2 and (b) P_T.

32. A 50-Ω resistor R_1 and a 30-Ω resistor R_2 are in parallel with a 15-Vdc source. Find (a) P_1 and P_2 and (b) P_T.

JOBS 7-7 to 7-9   Square Roots

Find the square roots of the following numbers.

1. 1936

2. 484

3. 121

4. 756.25

5. 302.76

6. 27,225

7. 45

8. 10.89

9. 219.04

10. 650.25

Perform the indicated operations.

11. $\sqrt{10^8}$

12. $\sqrt{10^{-8}}$

13. $\sqrt{64 \times 10^4}$

14. $\sqrt{360 \times 10^3}$

15. $\sqrt{810 \times 10^{-5}}$

16. $\sqrt{100 \times 10^{-4}}$

17. $\sqrt{0.04 \times 10^6}$

18. $\sqrt{12.1 \times 10^3}$

19. $\sqrt{6 \times 30 \times 10^6}$

20. $\sqrt{0.00625 \times 10^{-7}}$

JOBS 7–10 and 7–11 Applications of Exponential Power Formula

Solve each of the following.

1. A 100-Ω resistor dissipates 1 W of power. What is the current in the resistor?

2. A 1.5-kΩ resistor dissipates 300 mW of power. What is the current in the resistor?

3. How much voltage is across a 240-Ω resistor whose power dissipation is 60 W?

4. How much voltage is across a 50-kΩ resistor whose power dissipation is ¼ W?

5. What is the maximum current a 1-kΩ, ½-W resistor can carry without exceeding its power rating?

6. What is the maximum voltage a 220-Ω, 2-W resistor can handle without exceeding its wattage rating?

7. The power loss in a 200-ft length of No. 14 gage copper wire is 50 W. How much current is the wire carrying if the wire resistance is 0.5 Ω?

8. An 8-Ω load is dissipating 20 W of power. How much voltage is across the load?

9. A 50-Ω load is dissipating 100 W of power. How much current is flowing in the load?

10. How much is the resistance of a 75-W/120-V light bulb?

11. In Fig. 7-8, find P_1, P_2, P_3, and P_T.

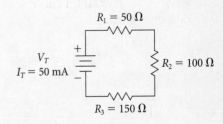

Figure 7-8

12. In Fig. 7-9, find P_1, P_2, P_3, and P_T.

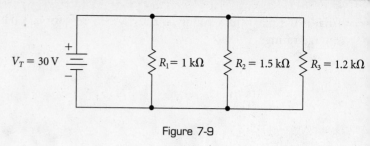

Figure 7-9

13. In Fig. 7-10, find V_T, I_1, I_2, I_3, I_T, P_1, P_3, and P_T.

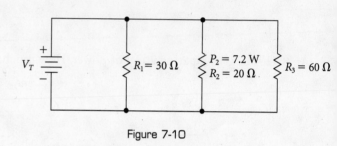

Figure 7-10

Algebra for
Complex Electric Circuits   8 chapter

| JOBS 8–1 and 8–2 | Combining Like and Unlike Terms |

Combine the following like terms.

1. $3b + 11b$

2. $10X + 6X + 2X$

3. $3y + 6y + 9y$

4. $15p - 1.5p$

5. $24c - 13c - 2.5c$

6. $20I + 6I + 2I - 8I$

7. $3a + 9a + 8a$

8. $12w - 62 + 100w - 42w$

9. $14\mu - 8\mu - \mu + \mu$

10. $8.5x - 2.5x + 8x$

11. $20 \text{ mA} + 6 \text{ mA} - 11 \text{ mA}$

12. $18I_2 - 6I_2 - 4I_2 + I_2$

13. $10a - 6a + 3a$

14. $4a - a + 6a - a$

15. $60C - 20C + 12C - 3C + 19C$

Combine the terms in the following algebraic expressions.

16. $4y + 5x + 7y + 13x$ 17. $4I + 2X + X + 2I$ 18. $13w - 6x - 2w + 8w + 13x$

19. $22a - 4p - 7a + 6p$ 20. $18R + 12 + 9R - 5$ 21. $66X - 14C + 3W + 8C - 36X + 19W$

22. $12p - 6b + 3b - 9p$ 23. $8a - 6x + 6x - 7a$ 24. $4I - I - 5 + 2I + 12$

25. $12a - 6x - 6a + 16x$ 26. $13p - 4a + 6p - 14a$ 27. $2.5x - 1.5a + 2.5a - 0.5a + 3x - 2x$

28. $18x - 32 + 2x - 4w$ 29. $10 - 4a + 3a - 6$ 30. $100 - 40I + 50 - 15I$

Solve each of the following.

31. Three resistors R_1, R_2, and R_3 are in series. If $R_2 = 2R_1$ and $R_3 = 3R_1$, express the total resistance R_T in terms of R_1.

32. Three resistors R_1, R_2, and R_3 are in series. If $R_1 = 5R_2$ and $R_3 = 2R_2$, express the total resistance R_T in terms of R_2.

33. Three resistors in parallel have branch currents designated I_1, I_2, and I_3. If $I_2 = 3I_1$ and $I_3 = 6I_1$, express I_T in terms of I_1.

34. Two resistors in parallel have branch currents designated I_1 and I_2. If $I_T = 4I_2$, express the value of I_1 in terms of I_2.

35. Two resistors R_1 and R_2 are in series. If $R_T = 3R_1$, express the value of R_2 in terms of R_1.

JOB 8-3 | Solving Simple Algebraic Equations

Solve the following equations.

1. $4X = 20$

2. $6R = 42$

3. $4X + 7X = 88$

4. $28 = 3a + 4a$

5. $\dfrac{P}{6} = 15$

6. $12X = 96$

7. $\dfrac{2a}{4} = 56$

8. $15b - 5 = 125$

9. $14I + 3I + I = 54$

10. $90 = \dfrac{3X}{4}$

11. $6a - 2a + 15 + 3a = 85$

12. $4X - 12 + X = 84$

13. $6R + 9 = 39$

14. $X + 2.5X + 1.5X + 4 = 99$

15. $7X + 1 = 64$

16. $7R + 8 = 64$

17. $X + 3.8X + 28 = 748$

18. $12a + a - 9 - 4a = 72$

19. $\dfrac{3b}{12} = 6$

20. $55 = 4.5L + 2.5L + 6$

21. $X + 1.5X = 7.5$

22. $3Q + 7Q - 18 + Q = 15$

23. $0.4X + 6 = 50$

24. $100 = 2M + 9M - M$

25. $300 = 6X + 9X - X + X$

Solve each of the following.

26. The total resistance R_T of three resistors R_1, R_2, and R_3 in series is 1200 Ω. If $R_2 = 3R_1$ and $R_3 = 400$ Ω, calculate the values of R_1 and R_2.

27. Two equal resistors (R) are connected in series with a 1000-Ω resistor to make a total resistance R_T of 2400 Ω. (a) Write an equation that can be used to solve for R. (b) Solve for R.

28. The emitter current I_E in a transistor circuit can be determined using the formula $I_E = I_B + \beta I_B$. Find the value of β if $I_E = 10.1$ mA and $I_B = 0.1$ mA.

29. The total current I_T in a three-branch parallel circuit is calculated as $I_T = I_1 + I_2 + I_3$. Find I_2 if $I_T = 10$ A, $I_3 = 4$ A, and $I_1 = 2I_2$.

30. Three resistors in parallel have branch currents designated I_A, I_B, and I_C. Find the value of each branch current if $I_B = 3I_A$, $I_C = 2I_B$, and $I_T = 24$ A.

JOBS 8–4 and 8–5   Solving Equations

Solve the following equations.

1. $\dfrac{9x}{6} + 8 = 20$

2. $\dfrac{Q}{4} - 25 = 40$

3. $15 = \dfrac{y}{4} + 13$

4. $24 = \dfrac{A}{6} - 18$

5. $\dfrac{4x}{9} - 18 = 30$

6. $\dfrac{3F}{9} + 80 = 100$

7. $17 = \dfrac{5x}{8} - 23$

8. $\dfrac{x}{21} + 3 = 24$

9. $119 - \dfrac{x}{4} = 33$

10. $12 = \dfrac{6c}{9} + 6$

11. $\dfrac{x}{6} - \dfrac{2}{3} = 8$

12. $2\dfrac{1}{5}x - 9 = 24$

13. $\dfrac{Q}{3} + 9 = 111$

14. $\dfrac{3x}{5} - 8 = 46$

15. $5\dfrac{2}{7}x + 4 = 115$

Combine the following terms.

16. $4a + 2 + a + 9$

17. $18x - 3 + 6 - 4x$

18. $8 - 12 + 3p - 5 - 9p$

19. $12y - 16 - 7y + 17 - 3y$

20. $4b + 20 - 8b - 15b + 12$

21. $-2x + 10 - 18x + 9 + 25x$

22. $-6 - 6c + 6 + 6c + 3$

23. $-13Z + 10Z + 15 - 35$

24. $a - b - 2 + 2b - 2a + 5$

25. $10x - 15 - 5x + 20$

Solve each of the following.

26. A radio and light in parallel draw a total current I_T of 1.5 A. The radio draws one-third as much current as the light. Find the current drawn by both the radio and the light.

27. A TV, VCR, and lamp in parallel draw a total current of 5.8 A. The TV draws 1½ times as much current as the VCR, and the lamp draws 1.2 A less than the TV. How much current is drawn by each of these items?

JOBS 8–6 and 8–7   Solving Equations with Signed Numbers

Solve the following equations.

1. $6x = 3x + 3$

2. $5a = 45 - 10a$

3. $6y + 48 = 18y$

4. $30 - 3I = 2I$

5. $2T = 4 - 6T$

6. $20x - 5 = 55 - 10x$

7. $16I - 2 = 6 - 8I$

8. $5R - 1 = -4R + 2$

9. $6x = 15x - 81$

10. $b + 36 = 90 - 8b$

11. $2x = 5x - 48$

12. $4x - 12 + 36 = 8x + 4x + 8$

13. $15y = 9y + 12.6$

14. $x = 5x - 10$

15. $3x - 15 = 2x + 5$

16. $-y = 9$

17. $5x = -15$

18. $24x = -72$

19. $9x + 51 = 15$

20. $-12 = -4R$

21. $5x + 2 = 2x - 1$

22. $-39 = 3x$

23. $13 = -3P + 4$

24. $3.2R + 20 = 1.2R + 6$

25. $0.8x = -48$

JOBS 8–8 to 8–10   Solving Equations Containing Parentheses

Remove parentheses and collect like terms.

1. $5(6 + 2a)$

2. $8(x + 4)$

3. $6(x - 5)$

4. $16 + 2(3 - 6x)$

5. $2(Q - 4) + 3(Q + 3)$

6. $45 - 3(5 - I_3)$

7. $4(3x - 4) - 5x$

8. $9I_1 - 5(I_1 + 2)$

9. $5y - 3(y - 2)$

10. $30 - 4(x + 6)$

Solve the following equations.

11. $5(x + 2) = 20$

12. $2(x - 3) = 16$

13. $3(Z + 6) = 21$

14. $3(3x + 2) + 5 = 65$

15. $10x - 2(x - 8) = 5x + 4$

16. $6a = 39 - (a - 3)$

17. $13x - (3 + 12x) = -3$

18. $9I_2 - (6I_2 - 2) = 27$

19. $-3(5 + F) = 12$

20. $40 = 5(V - 5)$

21. $\dfrac{x}{6} - \dfrac{x}{9} = \dfrac{2}{3}$

22. $\dfrac{V}{2} - \dfrac{V}{3} = 15$

23. $\dfrac{2x}{5} + \dfrac{x}{4} = 26$

24. $\dfrac{y}{3} - 21 = -\dfrac{y}{4}$

25. $\dfrac{a}{2} + \dfrac{a}{8} = 15$

Solve each of the following.

26. A transistor has an emitter current I_E of 12 mA and a beta β of 149. Find I_B if $I_E = I_B(\beta + 1)$.

27. The voltage gain A_V of a common-emitter amplifier is $A_V = \dfrac{R_C}{R_E + r'_e}$. Find R_E if

 $A_V = 10$, $R_C = 1.5$ kΩ, $r'_e = 5$ Ω.

28. The base input impedance Z_b of a common-emitter amplifier is calculated using the formula $Z_b = \beta(r'_e + R_E)$. Find r'_e if $Z_b = 153$ kΩ, $\beta = 150$, and $R_E = 1$ kΩ.

JOBS 8–11 to 8–13   Solving Simultaneous Equations

Solve the following sets of equation by either addition or substitution.

1. $x + 4y = 14$
   $x - 4y = -2$

2. $x + y = 8$
   $x - y = 2$

3. $9V + 2I = 34$
   $6V + 5I = -14$

4. $2a + b = 9$
   $4a - b = 6$

5. $5Z + 2R = 16$
   $3Z - R = 3$

6. $R_1 - 3R_2 = -8$
   $3R_1 + R_2 = 6$

7. $S + X = 0$
   $3S + 7X = 8$

8. $x + 2y = 1$
   $2x + 5y = 5$

9. $x - y = 0$
   $x + y = 2$

10. $5x + 2y = 36$
    $8x - 3y = -54$

11. $x - 5y = -5$
    $x + 3y = 3$

12. $8(y + 1) = 2x$
    $3(x - 3y) = 15$

123

Kirchhoff's Laws

9

c h a p t e r

JOBS 9–1 to 9–4	Kirchhoff's Laws

Answer each of the following.

1. Find the current flowing in I_5 of Fig. 9-1.

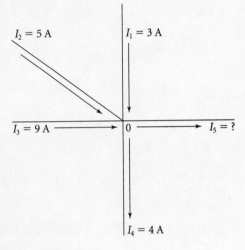

Figure 9-1

2. Determine (*a*) the current flow and (*b*) the voltage across each resistor in Fig. 9-2.

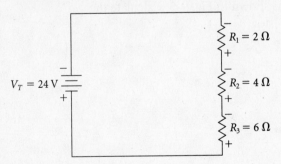

Figure 9-2

125

3. Find the current in the series circuit shown in Fig. 9-3.

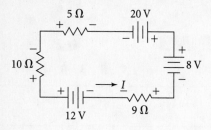

Figure 9-3

4. In Fig. 9-4, find the value of R_2 for the values given.

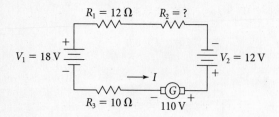

Figure 9-4

5. Two resistors are connected in parallel across a generator, as shown in Fig. 9-5. Calculate the generator voltage and the current flowing through each resistor.

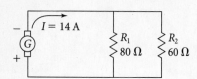

Figure 9-5

6. In Fig. 9-6, find all the missing values.

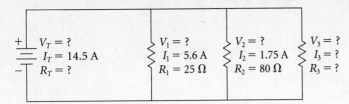

$V_T = ?$
$I_T = 14.5$ A
$R_T = ?$

$V_1 = ?$
$I_1 = 5.6$ A
$R_1 = 25\ \Omega$

$V_2 = ?$
$I_2 = 1.75$ A
$R_2 = 80\ \Omega$

$V_3 = ?$
$I_3 = ?$
$R_3 = ?$

Figure 9-6

7. Three resistors, one of 12 Ω, one of 24 Ω, and a third of unknown value are connected in parallel. The total resistance is 5.866 Ω. Find the value of the third resistor.

8. In the circuit given in Fig. 9-7, calculate the unknown current, the circuit voltages, and the total circuit resistance.

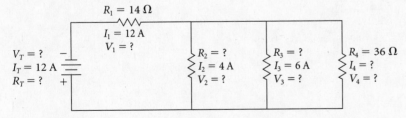

$R_1 = 14\ \Omega$
$I_1 = 12$ A
$V_1 = ?$

$V_T = ?$
$I_T = 12$ A
$R_T = ?$

$R_2 = ?$
$I_2 = 4$ A
$V_2 = ?$

$R_3 = ?$
$I_3 = 6$ A
$V_3 = ?$

$R_4 = 36\ \Omega$
$I_4 = ?$
$V_4 = ?$

Figure 9-7

JOBS 9–5 and 9–6   Using Kirchhoff's Law in Complex Circuits

1. In Fig. 9-8, find the values of I_L, I_1, and I_2.

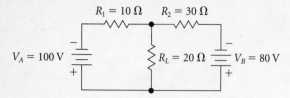

Figure 9-8

2. The circuit of Fig. 9-9 shows three unequal voltage sources. Calculate the current in each branch.

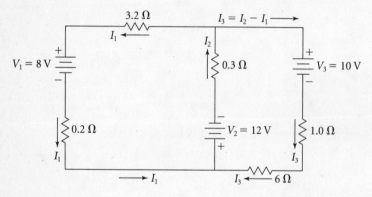

Figure 9-9

3. In Fig. 9-10, find the value of the current flowing through the 5-Ω resistor.

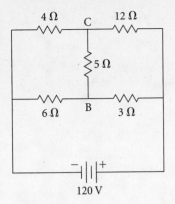

Figure 9-10

4. In the Wheatstone bridge circuit of Fig. 9-11, $R_1 = 1\ \Omega$, $R_2 = 1.5\ \Omega$, and $R_D = 3\ \Omega$. Find the value of R_x.

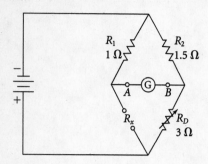

Figure 9-11

| JOB 9–7 | Equivalent Delta and Star Circuits |

1. In the delta circuit shown in Fig. 9-12, determine the resistances of the equivalent star circuit.

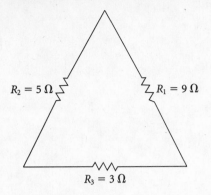

$R_2 = 5\,\Omega$ $R_1 = 9\,\Omega$

$R_3 = 3\,\Omega$

Figure 9-12

2. The three resistances in a delta-connected group shown in Fig. 9-13 have the following values: $X = 25\,\Omega$, $Y = 15\,\Omega$, and $Z = 40\,\Omega$. (*a*) Calculate the equivalent star circuit. (*b*) Using the three star resistances that you computed, show that when reconverted to a delta circuit, they will produce the values that were originally given.

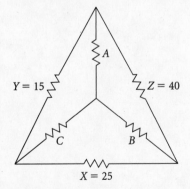

$Y = 15$ A $Z = 40$

C B

$X = 25$

Figure 9-13

JOBS 9–8 and 9–9   Thevenin's and Norton's Theorems

1. A 75-Ω load is to be connected to a two-terminal battery resistor network as shown in Fig. 9-14. Calculate the load current and the power delivered to the load.

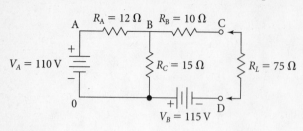

Figure 9-14

2. Using the data supplied in Fig. 9-15, determine the load current I_L by Norton's theorem.

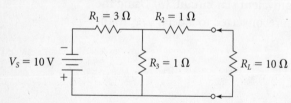

Figure 9-15

Applications of Series and Parallel Circuits

10

c h a p t e r

JOBS 10–1 and 10–2   Extended Range of Ammeters

1. Assume that you want to extend the range of a 1-mA 100-Ω meter movement to 10 mA (Fig. 10-1). Calculate the value of the shunt resistance required.

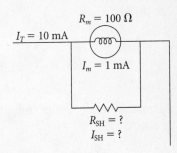

$R_m = 100\ \Omega$

$I_T = 10$ mA

$I_m = 1$ mA

$R_{SH} = ?$
$I_{SH} = ?$

Figure 10-1

2. Calculate the highest current that can be read by a 0–10-mA meter movement (Fig. 10-2) when a 4-Ω shunt is placed across it. The meter movement resistance is 100 Ω.

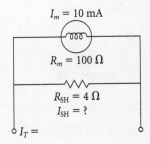

$I_m = 10$ mA

$R_m = 100\ \Omega$

$R_{SH} = 4\ \Omega$
$I_{SH} = ?$

$I_T =$

Figure 10-2

3. The ammeter shown in Fig. 10-3 is to measure a current of 100 mA. The meter movement sensitivity $I_m = 0.01$ A, and the shunt resistance is 0.50 Ω. Calculate the value of the internal meter resistance (R_m).

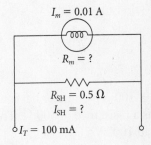

$I_m = 0.01$ A

$R_m = ?$

$R_{SH} = 0.5\ \Omega$
$I_{SH} = ?$

$I_T = 100$ mA

Figure 10-3

4. A meter movement has a resistance of 25 Ω, and it can carry a maximum current of 1 mA. Calculate the value of the shunt resistance required to measure 50 mA of current (Fig. 10-4).

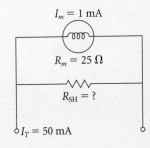

$I_m = 1$ mA

$R_m = 25\ \Omega$

$R_{SH} = ?$

$I_T = 50$ mA

Figure 10-4

JOBS 10–3 and 10–4   Voltmeters

1. Find the ohms per volt rating of a 35,000-Ω voltmeter that reads 25 V at full scale.

2. Calculate the multiplier resistance required in the circuit (Fig. 10-5) to limit the current flow to 1 mA.

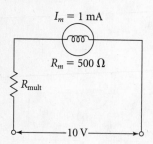

$I_m = 1$ mA

$R_m = 500\ \Omega$

R_{mult}

10 V

Figure 10-5

3. A 0–100-V voltmeter has a multiplier resistor of 99,500 Ω in series, with the moving coil that has a resistance of 500 Ω. The maximum current I_m is 1 mA. How can this circuit be altered to read 1000 V?

4. What current would be required for a 20,000 ohms-per-volt meter to obtain full-scale deflection (FSD)?

5. Calculate the voltage required to produce FSD when a 10,000-Ω resistor is connected in series with a meter movement rated at 1 mA and with an internal resistance of 100 Ω.

6. A 3-V voltmeter has a resistance of 150 Ω. Find the resistance of multiplier required so that the meter could read 75 V.

7. A voltmeter shows 10 V across a 100-Ω resistance. Find (a) the current through the resistor and (b) the value of multiplier needed for a voltmeter range of 100 V.

JOB 10–5   Voltage Dividers

1. The resistors in Fig. 10-6 are $R_1 = 5$ MΩ, $R_2 = 2$ MΩ, and $R_3 = 1$ MΩ. The voltage across AD is 30 V. Calculate the voltage drops across each resistor. The equivalent resistance R_e to R_2 and R_3 is $R_2 + R_3 = 3$ MΩ.

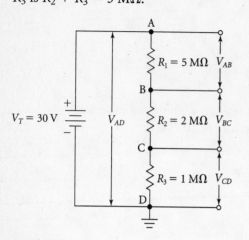

Figure 10-6

2. The voltage produced by the battery in Fig. 10-7 is 15 V. Resistor R_1 is 20 Ω. What resistance must resistor R_2 be in order to have a voltage drop across it of 6 V? (Verify your answer by computing the voltage drops and see whether they add up to V_T.)

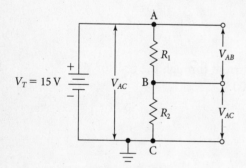

Figure 10-7

3. Design a voltage divider circuit for a 250-V power supply. The connected loads will draw currents of 20 mA, 30 mA, and 40 mA at 100 V. Allow a 10% bleeder current.

4. Find the values of the voltage divider resistors R_1, R_2, R_3, and R_4 in Fig. 10-8. The -50 bias terminal draws no current. The bleeder current is 10% of the total load current.

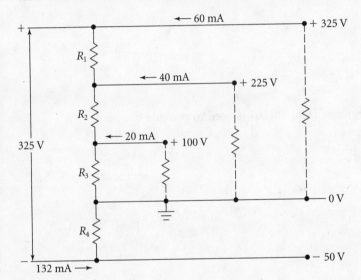

Figure 10-8

JOBS 10–6 to 10–8	Resistance Measurement

1. In Fig. 10-9, find the value of the unknown resistor R_a.

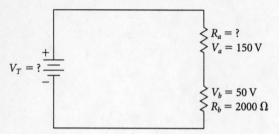

$V_T = ?$

$R_a = ?$
$V_a = 150\text{ V}$

$V_b = 50\text{ V}$
$R_b = 2000\ \Omega$

Figure 10-9

2. A voltage divider in the case circuit of a CE amplifier (Fig. 10-10) is used to provide the forward bias voltage. Find the voltage drop across R_2.

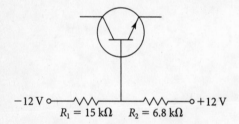

-12 V $+12\text{ V}$
$R_1 = 15\text{ k}\Omega$ $R_2 = 6.8\text{ k}\Omega$

Figure 10-10

3. In a balanced Wheatstone bridge circuit (Fig. 10-11), $R_1 = 20 \ \Omega$, $R_2 = 1000 \ \Omega$, and $R_3 = 100 \ \Omega$. Find the value of the unknown resistor R_x.

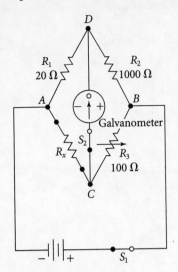

Figure 10-11

4. Figure 10-12 shows the voltage-dividing network of a power supply. Find (a) the Thevenin voltage V_{TH} for the load resistor R_L, (b) the Thevenin equivalent resistance, and (c) the load current I_L, assuming that $R_L = 550 \ \Omega$.

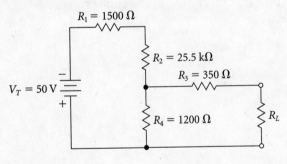

Figure 10-12

5. In the circuit of Fig. 10-13 find the current through R_L using Norton's theorem.

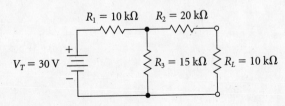

Figure 10-13

JOBS 10–9 and 10–10 | DC Equivalent Circuits for Self-Biased and Fixed Bias Transistor Circuits

1. In Fig. 10-14 find the value of (*a*) the load current I_{RL}, (*b*) the voltage V_{RL}, and (*c*) the collector voltage V_C.

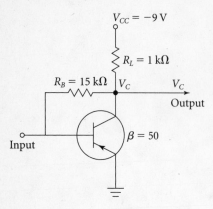

$V_{CC} = -9\,V$

$R_L = 1\,k\Omega$

$R_B = 15\,k\Omega$ V_C V_C

Output

$\beta = 50$

Input

Figure 10-14

2. In the fixed-bias circuit shown in Fig. 10-15, find (*a*) the collector current I_C, (*b*) the emitter current I_E, (*c*) the collector voltage V_C, and (*d*) the voltage across R_L.

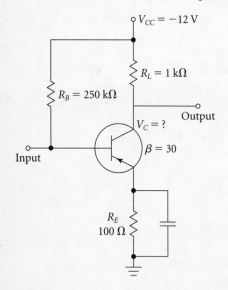

$V_{CC} = -12\,V$

$R_L = 1\,k\Omega$

$R_B = 250\,k\Omega$

$V_C = ?$

Output

$\beta = 30$

Input

R_E
$100\,\Omega$

Figure 10-15

1. Figure 10-16 depicts a common-emitter amplifier. The current flow in the collector is 1 mA. Find the values of (a) V_{CE}, (b) I_b, and (c) I_e.

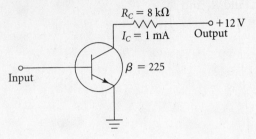

Figure 10-16

JOB 10–12   Attenuators

1. An L pad attenuator is inserted in a circuit to change the voltage delivered to the load R_L, shown in Fig. 10-17, without affecting the source current or source voltage. The load voltage is 10 V, the source voltage is 50 V, and the load resistance is 30 Ω. Find R_1 and R_2.

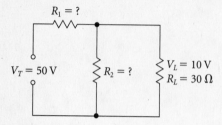

$R_1 = ?$

$V_T = 50\ V$ $R_2 = ?$ $V_L = 10\ V$
$R_L = 30\ \Omega$

Figure 10-17

2. A T-type attenuator is inserted in a circuit (Fig. 10-18) to reduce the voltage across R_L to 30 V. Find the values of R_1, R_2, and R_3 that will provide the required reduction in voltage.

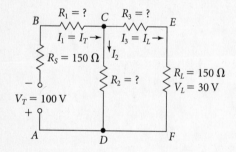

$R_1 = ?$ C $R_3 = ?$
B E
$I_1 = I_T \rightarrow$ $I_3 = I_L \rightarrow$
$R_S = 150\ \Omega$ I_2
     $R_L = 150\ \Omega$
$-$ $R_2 = ?$ $V_L = 30\ V$
$V_T = 100\ V$
$+$
A D F

Figure 10-18

JOB 10–13   Power and Resistor Size

Calculate the power size for the resistors in the following
circuits (Figs. 10-12 to 10-14, 10-17, and 10-18 are
repeated in this job for convenience).

1. In Fig. 10-12: P_L and P_3

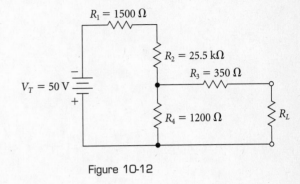

Figure 10-12

2. In Fig. 10-13: P_L and P_2

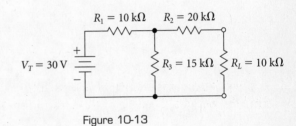

Figure 10-13

3. In Fig. 10-14: P_L

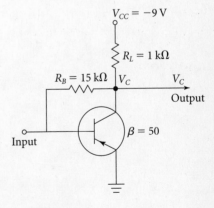

Figure 10-14

4. In Fig. 10-17: P_1 and P_2

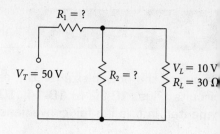

Figure 10-17

5. In Fig. 10-18: P_1, P_2, and P_3.

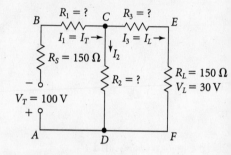

Figure 10-18

Efficiency 11
c h a p t e r

Percent

Write the following fractions as percents.

1. $\dfrac{82}{100}$

2. $\dfrac{3264}{10,000}$

3. $\dfrac{7}{1000}$

4. $\dfrac{1535}{1000}$

5. $3\dfrac{1}{7}$

6. $\dfrac{45}{67}$

7. $31\dfrac{5}{8}$

8. $\dfrac{467}{123}$

9. $\dfrac{23}{71}$

10. $\dfrac{249}{1137}$

Write the following percents as decimals.

11. 1.22%

12. 13.5%

13. 205.2%

14. 61.8%

15. 373%

16. 0.0081%

17. 0.51%

18. 143.5%

19. 7.04%

20. 0.003%

21. 0.025%

22. 15.08%

Write the following decimals as percent.

23. 55.02   24. 0.5   25. 1.07   26. 38.5   27. 0.27

28. 11.53   29. 0.0079   30. 0.037   31. 2.41   32. 103.7

33. 5.327   34. 201.3

Find what percent of the second number is represented by the first number in the following pairs of numbers.

35. 43, 17   36. 3, 18   37. 80/110   38. 4.6, 86   39. 2, 16

40. 120/45   41. 0.12/23   42. 18/20   43. 65/150   44. 0.7/36

Convert to a decimal and then reduce it to its lowest terms.

45. 80%   46. 120%   47. 340%   48. 56.7%   49. 0.1%

50. 720.5%   51. 2.7%   52. 42.8%   53. 0.03%   54. 0.63%

Find the value of the following.

55. 52% of 379   56. 17% of 458   57. 125% of 653   58. 0.035% of 725   59. 235.5% of 219

60. 2.25% of 175   61. 0.4% of 50   62. 33.4% of 115   63. 235% of 75   64. 5% of 83

Solve.

65. 30 is 20% of what number?   66. 18 is 55% of what number?   67. 45 is 15% of what number?

68. 88 is 20% of what number?   69. 2.7 is 5% of what number?   70. 43.5 is 15% of what number?

71. 27.4 is 25% of what number?   72. 105.6 is 40% of what number?

73. 11.3 is 18% of what number?   74. 4.7 is 0.5% of what number?

75. If a resistor color code shows that a resistor is 1500 Ω with a tolerance of ±5%. Find the range of minimum and maximum values for the resistor.

76. A resistor color code determines the value of resistance. The tolerance band gives the amount of deviation from the specified value. Calculate the amount of deviation of the following resistors.

    (*a*) 2.3 kΩ ±10%   (*b*) 6 MΩ ±5%   (*c*) 236 kΩ ±20%   (*d*) 4.5 MΩ ±5%   (*e*) 536.3 kΩ ±10%

    (*f*) 3.7 Ω ±1%   (*g*) 27.1 Ω ±20%   (*h*) 5.68 Ω ±1%   (*i*) 1.5 kΩ ±10%   (*j*) 23.7 kΩ ±5%

77. Find the average value of a sine wave that has a peak value of 146.5 V.

78. A voltmeter has a 5% accuracy rating. Find the range of readings for the following voltmeters.

    (*a*) 325 V min.– _____ max. _____   (*b*) 475 V min.– _____ max. _____

    (*c*) 36.5 V min.– _____ max. _____   (*d*) 2.07 V min.– _____ max. _____

Solve each of the following.

79. An apprentice is paid 65% of a technician's $695 per week. Calculate the apprentice's weekly income.

80. The local electronic supermarket is advertising a stereo system for $725 with no money down and just 12 easy payments of $75.25. Find (*a*) the amount of interest paid out over the 12 months and (*b*) the rate percent of simple interest that it represents.

81. A merchants buys a TV set for $368.30. He marks the price up 60% to cover his overhead. At a later date, he puts a sale sign on it: REDUCED 20%. What does the customer now pay for the TV set?

82. For the above question, calculate the sales tax of 8% and the final cost to the customer.

83. A customer borrows $500 from a bank and has to make 12 monthly payments of $58.95. What rate of interest does this represent?

84. Two resistors are connected in series across a 12.5-V supply. The voltage drop across one resistor is 5.3 V. What percent of the total supply voltage does this represent?

85. The FCC regulations state that a radio transmitter must stay within $\pm0.65\%$ of the assigned frequency on the FM band. Find the maximum and minimum allowable frequencies of a transmitted signal of 87.8 MHz.

86. An electrical circuit is fused with a 20-A fuse that is rated at 15% overload for a 4-s period. Find the amount of current that will flow through the fuse during this interval.

87. A computer circuit board contains 50 resistors, 27 capacitors, 14 transistors, and 17 ICs. Express each group of components as a percent of the total number of components.

88. A relay is designed to operate at ±10% of rated voltage. The line voltage is rated at 120 V, but the relay has only 106 V across it. Is this voltage within the normal operating range of the relay?

89. An electronic technologist earned $575 per week. His employer gave him a 15% raise. What was his weekly earnings after he received the increase?

90. An electric generator produces 220 V. The voltage drop in the line is 1.5%. Find the voltage drop in the line and the voltage supplied to the load.

91. An electrician has a 250-ft-long coil of BX cable. He uses 35 ft on an electrical job. What percent of the cable did he use?

92. Find the voltage produced by a generator if 97.5% of the generator voltage is delivered to a 120-V line.

93. An electric motor is rated at a no-load speed of 3380 rpm. However, under full load it rotates at 96% of its rated speed. Find the full-load speed.

94. A filter circuit is designed to have a cutoff frequency of 1675 Hz. Find 70.7% of this frequency.

95. In a bin of electrical wire weighing 53.4 kg, 17% was No. 22 wire, 12.3% was No. 18 wire, 39.6% was No. 24 wire, and the remainder was No. 14 wire. Find the weight of each type of wire.

96. Given a voltage with a peak value of 30 V. Find the (*a*) RMS value, (*b*) average value, and (*c*) peak-to-peak value.

JOB 11–4   Series Resonance

Change the following units of measurement.

1. 5.3 kW to watts

2. 3 hp to watts

3. 3200 W to kilowatts

4. 2 hp to kilowatts

5. ⅛ hp to watts

6. 5000 W to horsepower

7. 12 hp to watts

8. 39.83 kW to watts

9. 27.4 kW to horsepower

10. 1500 hp to kilowatts

11. 450 W to horsepower

12. 6.75 kW to horsepower

13. 5.7 hp to watts

14. 2¾ kW to horsepower

Solve each of the following.

15. Find the mechanical horsepower of a generator that delivers a current of 60.5 A at 230 V.

16. A diesel engine delivers 1450 hp. Find its electrical rating in kilowatts.

17. A motor consumes 6840 W of electricity. What value of mechanical horsepower does this represent?

18. An electric motor delivers 10.5 hp. Find the equivalent electrical watts.

19. A generator delivers 1¼ kW. Find the equivalent horsepower rating.

20. A shunt motor delivers 6.9 hp. Find the watt output of the motor.

JOBS 11–5 and 11–6   Electric Devices

Solve each of the following

1. An antenna motor rated at 12 hp consumes 9.23 kW. Find the efficiency of the motor.

2. An electric motor consumes 3640 W. Find the horsepower it can deliver if its efficiency is 85%.

3. A motor delivering 5.5 hp draws 21.3 A from a 230-V line. Find (*a*) how much electrical power the motor is taking from the line and (*b*) the efficiency of the motor.

4. A motor delivers 10.5 hp and consumes 8.5 kW. Find the efficiency of the motor.

5. A 30-hp motor drives a generator that delivers 45.5 A at 440 V. Find the efficiency of the generator.

6. A dc motor receives 8.98 kW from the line and delivers 10 hp to the load. Find the efficiency of the motor.

7. A dc generator receives 18.6 kW of energy and delivers 74.6 A at 230 V. Find the efficiency of the generator.

8. Find the efficiency of a transmission line when the resistance of the lines are 0.4 Ω. The generator delivers 130 V to a load that draws a current of 18 A.

9. A dc motor delivers 20 hp at full load and receives 15.6 kW from the line. Find the efficiency of the motor.

10. A dc generator receives 20 kW of energy and delivers 34.6 A at 440 V at full load. Find the efficiency of the generator.

11. Find the efficiency of a dc generator being driven by a 12.5-hp motor, and delivering 68.9 A at 220 V.

12. It requires 30 hp to drive a generator that delivers 45.5 A at 440 V. Find the efficiency of the generator.

13. A power transformer draws 75 W and delivers 65 W to a receiver. Find its efficiency.

14. A generator produces ¾ kW and receives 2.5 hp. What is its efficiency?

15. A power transformer draws 2.5 A from a 130-V line. Find the efficiency of the transformer if it delivers 250 V at 0.8 A.

16. An electric motor rated at 10.5 hp has an efficiency of 85%. Find the number of kilowatts of energy it consumes.

17. A generator that is 90% efficient delivers 11 kW. What horsepower rating must a diesel engine have to drive the generator?

18. Find the electrical horsepower of a generator that delivers 55 A at 230 V.

19. A 3-hp motor is connected to a 220-V line. How much current does the motor take from the line?

20. A 9.5-hp motor has an efficiency of 85%. How many kilowatts are required to drive the motor?

21. A transmission line receives 15 kW from a generator and delivers 12.8 kW to a motor. Find the efficiency of the transmission line.

22. A motor operates at an efficiency of 95% and draws 7.3 A from a 120-V line. Find the horsepower output.

23. Find the kilowatt input of a motor that delivers 12 hp to a load at 87% efficiency.

24. Find the horsepower output of a motor that has an efficiency of 89% at full load. The motor receives 32.6 A at 440 V.

25. Find the kilowatt input to a motor that delivers 12 hp to its load at 80% efficiency.

26. Find the output of a dc motor in horsepower. The motor has an efficiency of 92% at full load and receives 69.25 A at 230 V from the line.

27. Calculate the kilowatt input of a 110-hp motor that has an efficiency of 89.5%.

28. A dc motor draws 53.4 A at 130 V. If the motor has an efficiency of 78.5%, find the horse-power output of the motor.

29. A power transformer that has an efficiency of 85% delivers 75 W. Calculate the input power to the transformer.

30. A 230-V 7-hp motor has an efficiency of 80%. Find (*a*) the input power in kilowatts and (*b*) the current that it draws from the line.

31. An electric generator has an internal impedance of 3 Ω. The open circuit voltage is 15 V. Find (*a*) the load resistance that will receive the maximum power and (*b*) the maximum power.

32. Find the maximum power that can be delivered by a 25-V generator that has an internal resistance of 4 Ω.

Resistance of Wire

12 chapter

| JOBS 12-1 to 12-3 | Ratio and Proportion |

Express the following quantities as a ratio in simplest form.

1. $\dfrac{15}{20}$

2. $\dfrac{1}{3} : \dfrac{1}{5}$

3. $15:3$

4. $27:3:9$

5. $8:12:16$

6. $\dfrac{1}{4} : \dfrac{3}{8}$

Determine which of the ratios in each of the following pairs is larger.

7. $\dfrac{5}{6}$ and $\dfrac{11}{12}$

8. $2:5$ and $4:9$

9. $2:3$ and $3:4$

10. $3:5$ and $7:9$

11. $9:10$ and $11:12$

12. $\dfrac{5}{7}$ and $\dfrac{3}{4}$

Write each of the following ratios as a fraction with the denominator 60.

13. $3:5$

14. $\dfrac{4}{15}$

15. $7:12$

16. Divide 153 into 3 parts in the ratio of $2:3:4$.

17. The ratio of the length and width of a rectangular field is $5:3$. Find the dimensions if the area is 60 ft².

18. A transformer has 720 turns of wire on the primary and 80 turns on the secondary. What is the turns ratio of primary to secondary?

19. A large wheel with a diameter of 15 in drives a smaller one that has a diameter of 3 in. Find the mechanical advantage.

20. A 375-ft length of copper wire has a resistance of 0.9465 Ω. Find the length of a copper wire of the same area that has a resistance of 12.624 Ω.

21. A wire cable 118 ft long costs $12.35. What would 325 ft cost at the same rate?

22. A copper wire whose resistance is 6.257 Ω has a diameter of 33.427 mils. Find the resistance of a wire that is the same material and length and has a diameter of 47.602 mils.

23. What is the ratio of 2 hp to 2 kW?

24. Calculate the efficiency of a motor that has an output of 2.5 hp and an input of 2.2 kW.

25. Calculate the current gain (β) of a transistor that has the following values: $I_\beta = 0.06$ mA and $I_C = 3$ mA.

26. The formula for finding the power factor (PF) of an ac circuit is the ratio of its resistance R to its impedance Z. Given a circuit with a resistance of 17.5 Ω and impedance of 105 Ω, find the PF (express it as a percent).

27. The ratio of a coil's reactance to its resistance is called the coil's Q factor. In a given circuit, the reactance of a coil is 3300 Ω and the resistance is 75 Ω. Find the Q of the coil.

28. In a Wheatstone bridge circuit $R_1 = 25\ \Omega$, $R_4 = 42\ \Omega$, and $R_2 = 250\ \Omega$. R_x is the unknown

    resistance. Find the value of R_x in the formula $\dfrac{R_x}{R_2} = \dfrac{R_1}{R_3}$.

29. A coil and resistor are connected in series. The voltage drop across the 150-Ω resistance is 23 V. Find the resistance of the coil, which has a voltage drop across it of 9 V.

30. Find (*a*) the fourth proportional to 3, 6, and 12 and (*b*) the third proportional to ¼ and ⅙.

31. A cement mixture contains 5 parts of sand, 3 parts of gravel, and 2 parts of cement. Find the weight in kilograms of each when the total weight is 108 kg.

32. A copper wire 10 m long has a resistance of 12.3 Ω. Find the resistance of 21 m of the same wire.

33. A young boy is standing beside a flagpole, which casts a shadow of 15 ft. The boy, whose height is 4.5 ft, casts a shadow of 3.7 ft. Find the height of the flagpole.

34. A metal alloy is composed of copper and zinc. In one alloy the ratio of the weight of copper to that of zinc is $19:11$. How many kilograms of copper should be used to make 180 kg of the alloy?

35. Two resistors R_1 and R_2 are connected in parallel. R_1 is 75 Ω, and R_2 is 150 Ω. The current through R_1 is 0.5A. Find the current flowing through R_2.

36. A conductor has a resistance of 64 Ω and a cross-sectional area of 30 mils. Another conductor of the same length and material has a cross-sectional area of 110 mils. Find the resistance of the second conductor.

37. A gear having 12 teeth meshes with one having 26 teeth. If the larger gear travels at 500 rpm, at what speed should the small gear be driven?

JOB 12–4   The American Wire Gage Table

Find the missing values in the following problems without using the AWG table.

Problem	Gage No.	Diameter (mil)	Diameter (in)	Area (cm)
1.		10.03		
2.	11			
3.				320.4
4.			0.323	
5.	15			
6.			0.0453	
7.	0000			
8.				133,100
9.		5.615		
10.				83,690

JOBS 12–5 to 12–10	Resistance of Wires

1. A nichrome wire has a resistance of 65 Ω. Find the resistance of an iron wire of the same length and diameter.

2. Find the resistance of a silver wire that has the same length and diameter as a copper wire that has a resistance of 17 Ω.

3. A copper wire that is 1500 ft long, has 10.58 A flowing through it when there is a voltage drop of 20 V across the ends of the wire. Find the gage of the wire.

4. A copper wire has a resistance of 35 Ω and a diameter of 50.82 mils. Find the resistance of a wire of the same length but an area of 100.5 c mil.

5. An electric heater uses 14 ft of nichrome wire that has a diameter of 0.008 in. Find (*a*) the resistance of the nichrome wire, (*b*) the amount of current it will draw from a 115 V line, and (*c*) the amount of power it will consume.

6. A coil of copper wire has a resistance of 2.54 Ω at a temperature of 20°C. The diameter of the wire is 0.064 in. Find the length of the wire.

7. How many feet of annunciator copper wire that has a diameter of 0.054 in. are required to make a resistance of 3 Ω?

8. A wire-wound resistor is made with manganin wire 0.04 in. in diameter. Find the length of wire needed to form a 25-Ω coil.

9. A rheostat of 45 Ω is made with 350 ft of German silver wire. What gage of wire should be used?

10. A nichrome wire 11 ft long has a resistance of 25.2 Ω. Find the diameter of the wire in mils.

JOB 12-11   Resistance of Wires

1. A coil made of copper wire has a resistance of 45 Ω at 20°C. What is its resistance at 80°C?

2. A resistance thermometer is made of copper wire. Its temperature is 20°C, and its resistance is 55 Ω. How much will its resistance increase for a temperature rise of 2°?

3. The field windings of a shunt motor have a resistance of 100 Ω at 25°C. After the motor runs for a few hours, the resistance of the field windings is 160 Ω. Find the temperature of the coils.

Size of Wiring

13

c h a p t e r

| JOB 13-1 | Maximum Current-Carrying Capacities of Wires |

Solve each of the following.

1. Find the resistance of 2000 ft of No. 0 wire.

2. Find the resistance of 500 ft of No. 14 wire.

3. What size of wire will have a resistance of 1.32 Ω per 1000 ft?

4. What gage of wire should be used if the circular-mil area required is 95,680 cmils?

5. A No. 12 copper wire has a diameter of 80.8 mils. Find the resistance of 13,500 ft.

6. What is the resistance of a copper wire that is 1200 ft long and has a diameter of 257.6 mils?

7. A copper wire has a resistance of 9.21 Ω. The diameter is 64 mils. Find the length of the wire.

8. Find the resistance of 775 ft of No. 12 wire supplying the current for a battery charger that operates an emergency lighting system.

9. A wire 3525 ft long and 87 mils in diameter has a resistance of 0.473. Find the resistance of a wire of the same material that is 4736 ft long.

10. A copper wire 1763 ft long has a resistance of 0.78 Ω. Find the resistance of the same wire that is 1 mi long.

11. An engineer is designing a shunt for a milliammeter. The shunt must have a resistance of 2.76 Ω. A No. 44 enameled copper wire that has a resistance of 4.34 Ω/1000 ft is available. Find the length of wire required.

12. The resistance of 90.4 ft of wire is found to be 23.4 Ω. A coil of wire made of identical wire has a resistance of 814 Ω. Find the length of this wire.

13. What is the maximum current that may be safely carried by No. 10 asbestos-insulated wire?

14. Find the voltage drop in a 1000-ft length of No. 10 varnished-cambric-insulated wire when it is carrying its maximum current.

15. What is the smallest rubber-covered wire that can carry 90 A safely?

16. What is the resistance per foot of the smallest varnished-cambric-insulated wire that can carry 75 A safely?

17. A coil of copper wire has a resistance of 0.792. If the diameter is 80.8 mils, find the length of the wire.

JOB 13-2   Finding the Minimum Size of Wire to Supply a Given Load

Solve each of the following.

1. What is the minimum size of varnished-cambric-insulated wire that may be used to supply a load of 35 lamps, each of which draws 0.8 A?

2. Four electric heaters, each drawing 10 A of current, are connected in parallel to a 120-V circuit. Find the minimum size of rubber-covered wire that may be used to supply power to the load.

3. What is the minimum size of asbestos-covered wire that may be used to supply a load of thirty 200-W lamps, twenty-five 100-W lamps, and thirty-five 60-W lamps, all connected in parallel across a 115-V source?

4. A 5-, 10-, and 12-hp motor are all to be connected to a 440-V line. Find the minimum size of rubber-covered wire that may be used to wire the circuit.

5. A 15-hp motor that has an efficiency of 85% is to be connected to a 130-V line. Find the smallest size of varnished-cambric-covered wire that may be used.

6. What is the minimum size of RH wire that can supply an 8.2-Ω load at 220 V?

7. A 120-V line supplies power to seven computer stations. Each computer is rated at 150 W, and each monitor is rated at 350 W. What size of RH wire is required to supply power to the seven stations?

8. An electric stove rated at 35 kW is connected to a 220-V line. Find the smallest size of type RH wire that can be used to supply power to the stove.

9. A 25-hp motor is to be connected to a 220-V source. The input to the motor must be equal to 112% of the rated horsepower. Find (*a*) the power to be delivered to the motor and (*b*) the minimum size of varnished-cambric-insulated wire that must be used.

10. An electric heater has a resistance of 1.6 Ω. What is the minimum size of RH wire that is required to connect the heater to a 120-V line?

| JOBS 13–3 and 13–4 | Size of Wiring |

Solve each of the following.

1. Ten kilowatts of power is to be transmitted 2640 ft from a generator that maintains a constant potential of 220 V. The allowable line drop is 10%. Find the size of wire that must be used.

2. A load of 15 A is 400 ft from a 120-V source. If No. 10 wire is used, find (*a*) the voltage drop and (*b*) the voltage at the load.

3. A motor is 1500 ft from a generator that produces a current of 25 A. A copper wire that has a diameter of 128.5 mils is used to connect the two. Find the voltage drop in the line.

4. A load is 1200 ft away from the source. The current in the circuit is 25 A, and the voltage across the load is 220 V. Voltage drop cannot exceed 2%. Find the size of copper conductor that is required.

5. A 12-hp, 220-V motor is located 850 ft from a power source that maintains a constant potential of 232 V. The efficiency of the motor is 83%. What size of wire should be used to connect the motor to the source?

6. A generator with brush voltage of 120 V is located 400 ft from a motor that requires 20 A at 110 V. What size of wire must be used to connect the generator and the motor?

7. Find the smallest size of wire (type RH) that will conduct a current to 75 lamps, each with a resistance of 270 Ω, at a distance of 350 ft from a source producing 115 V. The lamps are connected in parallel and have 108 V across the load.

8. A 220-V source supplies power to a 35-kW load at a distance of 125 ft. The allowable voltage drop is 5%. Find the smallest size of varnished-cambric-insulated wire that may be used.

9. A 120-V generator is located 560 ft from a building. The load within the building has 450 lamps, each drawing 0.15 A. The efficiency of the transmission line must be 90%. What size of copper wire must be used?

10. A pump is located some distance from the power source. It draws 5.5 A at 120 V. The conductor used is No. 8 wire, and its line drop is 2% of the motor voltage. Find the maximum length of cable that can be used (refer to Table 13-1 in the textbook to obtain resistance per foot).

Trigonometry for Alternating- **14**
Current Electricity
c h a p t e r

JOBS 14–1 and 14–2	Using Trigonometric Functions

Answer each of the following.

1. Find the sine, cosine, and tangent of $\angle X$ in the triangle XYZ shown in Fig. 14-1.

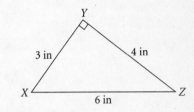

Figure 14-1

2. Find the sine, cosine, and tangent of $\angle D$ in the triangle DEF shown in Fig. 14-2.

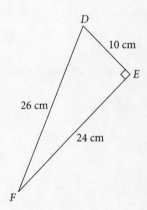

Figure 14-2

3. Calculate the sine, cosine, and tangent of the angles named in each triangle in Fig. 14-3.

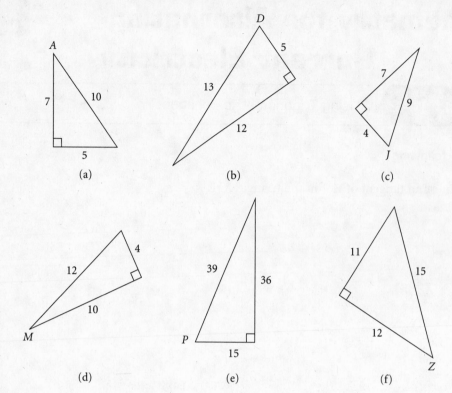

Figure 14-3

Find the value of the following.

4. Sin 17° =

5. Cos 47° =

6. Tan 85° =

7. Sin 63° =

8. Tan 32° =

9. Cos 18° =

10. Sin 37° =

11. Cos 71° =

12. Tan 4° =

Find the number of degrees in each angle.

13. Sin A = 0.7771

14. Cos B = 0.9945

15. Tan C = 19.081

16. Sin D = 0.1219

17. Cos E = 0.5736

18. Tan F = 0.2679

19. Sin G = 0.9744

20. Cos H = 0.7547

21. Tan I = 1.1106

Find the number of degrees in each angle, correct to the nearest degree.

22. Sin 0.4430

23. Cos 0.0935

24. Tan 0.1840

25. Sin 0.8706

26. Cos 0.9855

27. Tan 10.495

28. Sin 0.9867

29. Cos 0.6109

30. Tan 1.063

JOB 14-3   Finding the Acute Angles of a Right Triangle

Find ∠A and ∠B in the right triangle shown in Fig. 14-4 if:

1. AC = 18 ft and BC = 28 ft

2. AB = 25.2 in and BC = 20.5 in

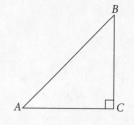

Figure 14-4

3. BC = 24 W and AB = 36 W

4. AC = 22 m and BC = 60 m

5. AB = 32.35 ft and BC = 12.75 ft

6. AB = 48 lb and AC = 20 lb

Solve each of the following.

7. Calculate the angle between the roof and the ceiling joists in Fig. 14-5. The rise is 2.75 m, and the run is 5.67 m.

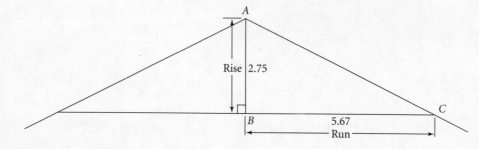

Figure 14-5

8. A ski lift makes an angle of 25° with the ground. The ski lift is 2275 ft long. How high is the mountain?

9. From an airplane flying at 15,750 m the angle of *depression* to an airport is 28°. Assuming the terrain to be horizontal between the airport and the spot below the airplane, find the horizontal distance from the airplane to the airport.

10. In a resistive-inductive circuit, the phase angle between the total voltage V_T and the resistive voltage V_R is 47° (Fig. 14-6). The voltage across V_R is 90° out of phase with the voltage across the inductor V_L. The voltage across V_R is 55 V. Find (a) the total voltage V_T and (b) the voltage across the inductor V_L.

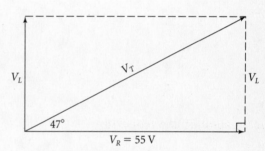

Figure 14-6

11. For the machine tool shown in Fig. 14-7, solve for dimension A.

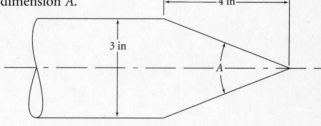

Figure 14-7

12. A vertical pole 15 ft high casts a shadow 22 ft 4 in long. Find the angle of elevation to the top of the pole.

13. The phase angle between the true power of an alternator and the apparent power of the alternator is 21°. The true power is 147.5 W. What is the apparent power? (Refer to Fig. 14-8.)

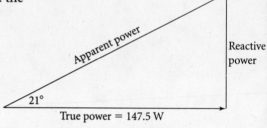

Figure 14-8

| JOBS 14–4 and 14–5 | Trigonometry |

Use the triangle shown in Fig. 14-9 to solve the following.

1. Find AC and $\angle B$ if $AB = 32$ m and $\angle A = 72°$

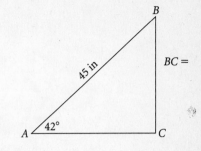

Figure 14-9

2. Find BC and $\angle B$ if $AB = 42$ in and $\angle A = 35°$

3. Find AC and $\angle A$ if $BC = 82$ cm and $\angle B = 45°$

4. Find AC and $\angle B$ if $AB = 350$ V and $\angle A = 30°$

5. Find AB and $\angle B$ if $BC = 85$ W and $\angle A = 28°$

6. Find BC and $\angle A$ if $AC = 450$ ft and $\angle B = 35°$

7. Find AC and $\angle A$ if $AB = 105$ in and $\angle B = 18°$

8. Find BC and $\angle B$ if $AB = 385$ V and $\angle A = 23°$

Solve each of the following.

9. In triangle *ABC* (Fig. 14-10), find (*a*) side *BC* and (*b*) angle *B*.

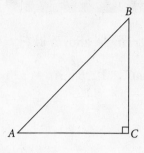

Figure 14-10

10. In a resistive-capacitive circuit, the voltage across the resistor V_R is 175 V, and the voltage across the capacitor V_C is 83 V (Fig. 14-11). Find (*a*) the total applied voltage V_T and (*b*) the phase angle θ.

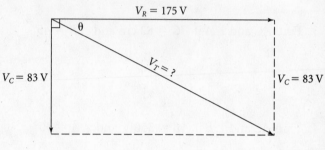

Figure 14-11

11. The impedance *Z* of a resistive inductive circuit (Fig. 14-12) is 43.7 kΩ. The resistance *R* is 21.4 kΩ. Find the (*a*) inductive reactance X_L and (*b*) phase angle θ.

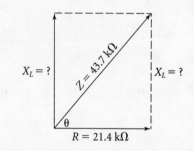

Figure 14-12

12. In a given rectangle the length is 87 in and the width is 43 in. Find the angle that the diagonal makes with the longer side.

13. Calculate the spacing for eight equally spaced holes on the circle with a radius of 6 in (Fig. 14-13). This circle is called a *bolt circle*, because bolts are to fit through the holes after they are drilled. Draw the altitude *OC*, which divides the line *AB* into two equal parts.

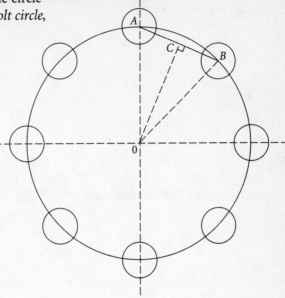

Figure 14-13

14. A chord of a 10-in circle makes an angle of 30° with the diameter at the circumference. Calculate the length of the chord.

15. A cone has an altitude of 24 in and a slant height of 28 in. Calculate the size of its vertical angle.

16. A small boat is observed from the top of a cliff 250 ft above sea level at an angle of depression of 15°. How far is the boat from the base of the cliff?

17. The treads of a staircase are 12 in, and the rise of each stop is 8 in. Find the angle of elevation of the stairs.

18. A blueprint (Fig. 14-14) shows how a tinsmith should bend a piece of metal to an angle of 110°. The end piece is 120 cm. Find the distance X that signifies how far back the bent piece is going to reach.

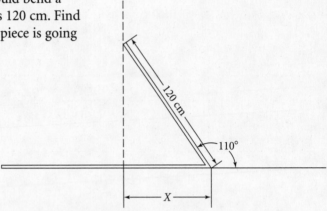

Figure 14-14

Introduction to AC Electricity

15 c h a p t e r

JOB 15–1	Graphs

Plot the graph for each of the following.

1. The distances and times required to reach certain points on a trip.

	A	B	C	D	E
Distance (miles)	42	105	130	185	205
Time (hours)	1	2.25	3	4.25	5

2. Population of Sweden.

Years	1871	1881	1891	1901	1911	1921	1931	1941	1951	1961
Population (million)	3.5	4.1	4.5	5.3	6.4	8.5	10.1	11.8	14	18.1

3. Permeability (B-H) curve for mild steel.

H (amp/meter)	100	200	300	400	500	600	700	800	900	1000	1100	1200	1300	1400	1500
B (tesla)	0.1	0.4	0.75	9.5	1.08	1.2	1.3	1.35	1.38	1.4	1.45	1.47	1.48	1.49	1.5

4. Output power versus load resistance.

Load resistance (ohms)	0	0.25	0.5	0.75	1.0	1.25	1.5	1.75	2.0
Power at load (watts)	0	23	33	35	36	35	34	33.1	32.5

5. Time constant discharge curve for *RC* circuit. The capacitor discharges 63% in each time constant.

Time constant	0	1	2	3	4	5
Percent of voltage	100	37	13.7	5.1	1.9	0

JOBS 15–2 to 15–4	AC Waves

Solve each of the following.

1. What is the wavelength of an AM radio station that broadcasts at a frequency of 610 kHz?

2. What is the wavelength of an FM radio station that broadcasts at a frequency of 88.7 MHz?

3. What is the frequency of a radio wave if its wavelength is 185 m?

4. A sine wave of voltage is applied to a resistor and has a maximum value of 113 V. Find the instantaneous value at 75°.

5. A sine wave has an effective value of 90 mA. Find the instantaneous value at 65°.

6. An ac voltage has an instantaneous value of 102 V at 42°. Find the maximum value of the wave.

7. Find the phase angle at which an instantaneous voltage of 56 V appears in a wave that has a maximum value of 120 V.

8. A sine wave voltage has an effective value of 105 V. What is its maximum value?

9. What is the effective value of an alternating current with a peak value of 8 A?

10. A maximum ac voltage of 160 V produces a peak current of 8 A through a resistive circuit. Determine the effective ac voltage and current.

JOBS 15–5 and 15–6	Alternating Current

Solve each of the following.

1. Find the frequency of a sine wave having a period of (*a*) 112 ns, (*b*) 19.2 μs, and (*c*) 10.2 ms.

2. What is the period of a 7-kHz wave?

3. If the gain setting on an oscilloscope is 15 V per division, what is the peak voltage of the signal?

4. The gain is 100 mV per division, and the time base is 200 ms per division. Find (*a*) the amplitude and (*b*) the frequency of the signal shown in Fig. 15-1.

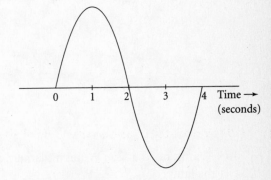

Figure 15-1

5. If the signal in Fig. 15-1 is a standard 4-V peak-to-peak, what is the scale of the vertical amplifier?

6. If the signal in Fig. 15-1 is 5 kHz, what is the horizontal scale?

For questions 7 to 10, draw a phasor diagram (not to scale).

7. A 75-Ω resistor rated at 15 W is connected to 120 V. Find the current flowing through the resistor.

8. Find the voltage needed to operate a 1200-W toaster that draws 11.5 A of current. What is its resistance?

9. Find the voltage needed to operate a 1500-W neon sign that has a resistance of 75 Ω.

10. A lamp rated at 250 W has a filament resistance of 72.8 Ω. Find the current flowing through the lamp.

JOB 15–7	The Pythagorean Theorem

Find the unknown side in each of the following right triangles.

Problem	a	b	h
1.	?	8	10
2.	9	?	14
3.	8.2	?	16.3
4.	56	71.7	?

Solve each of the following.

5. A piece of steel is 25 cm long and 9 cm wide. Find the length of the diagonal, and correct to one decimal place.

6. A baseball diamond is a square with sides 90 ft long. Find the distance from home plate to second base.

7. Find the length of a common rafter used on a roof of a house that has a rise of 22.5 ft and a run of 16.25 ft.

8. Figure 15-2 shows a house roof design. Rafters $XW = 29.2$ ft, and rafter $XZ = 11.5$ ft. Find the length of XY, YZ, and ZW (angles of YXW and XZY are right angles). *Hint:* For the second part, compare the ratios of the corresponding sides of similar triangles XZW and XYZ to find the length of YZ before going on to complete the solution.

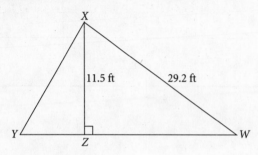

Figure 15-2

JOB 15-8	Power

Draw a phasor diagram for each problem (not to scale).

1. A coil of 0.4 H is connected to a 120-V source at a frequency of 2 kHz. Find the current flow.

2. What voltage is required to produce a current flow of 0.35 A through an inductance of 350 mH at a frequency of 250 Hz?

3. Find the effective power consumed by the 350-mH coil in question 2.

4. Find the reactance of a coil that draws 125 mA of current when connected to a 220-V source. What is the inductance of the coil if the frequency is 500 Hz?

5. A coil connected to a 120-Hz source has a reactance of 4.5 kΩ and a resistance of 500 Ω. What is the power factor?

6. Find the current through the coil in question 5 when the coil is connected to 110-V source.

7. How much power is dissipated in the coil of question 6?

8. Find the phase angle between the voltage and current for the coil in question 5.

JOB 15–9   Harmonic Analysis

Answer each of the following.

1. A voltage signal is said to be composed of two groups of signals. Name the two groups.

2. State two reasons why harmonics are important in modern-day electronic components.

Inductance and Transformers

16 chapter

JOB 16-1   Inductance

Define the following.

1. MMF

2. Reluctance

3. Permeability

4. Flux density

5. Magnetizing force

Answer the following.

6. What is the difference between MMF and magnetizing force?

JOB 16–2   Reactance of a Coil

Answer each of the following.

1. Define inductance.

2. State the two important factors that affect the amount of reactance in a coil.

3. Calculate the X_L of a 150-mH coil at (*a*) 1 MHz and (*b*) 15 MHz.

4. What is the inductance of a coil that has a reactance of 1500 Ω at a frequency of 50 kHz?

5. Find the X_L of a coil with negligible resistance that is connected to 110 V with 0.05 A of current.

6. At what frequency will an inductance of 250 mH have the reactance of 1500 Ω?

7. A 25-mH coil is operating at a frequency of 1 kHz. Find (*a*) the inductive reactance and (*b*) the current flow if the voltage across the coil is 115 V.

8. What must be the inductance of an RF choke if it is to have a reactance of 1250 Ω at 60 Hz?

9. A coil in the tuning circuit of a radio transmitter has an inductance of 125 mH. The transmitter is operating at a frequency of 6.1 kHz. Find (*a*) X_L and (*b*) the current when the voltage across the coil is 175 V.

JOBS 16–3 to 16–6   Coils and Inductance

Solve each of the following.

1. A 5-H coil and a 1500-Ω resistor are connected in series across a 110-V 60-Hz line. Find (a) X_L, (b) Z, (c) I_T, (d) V_R and V_L, (e) θ, (f) PF, and (g) power used.

2. A series circuit consists of a 1200-Ω resistor and a 7-H choke. The voltage drop across the resistor is 70 V at 60 Hz. Find (a) X_L, (b) Z, (c) I, (d) V_L, (e) V_T, (f) θ, (g) PF, and (h) power used.

3. How much is the impedance Z of a coil that allows 150 mA when connected to 115-V 60-Hz source? The resistance of the coil is 7.5 Ω.

4. A coil has a resistance of 7 Ω and an inductive reactance of 17 Ω at a certain frequency. Find (a) the Q of the coil and (b) the Z of the coil.

5. A 275-μH inductance has a Q of 27 at 1.1 Hz. Find the resistance of the coil.

6. What is the inductance of a coil that draws 45 mA from a 110-V 60-Hz source? The resistance of the coil is 30 Ω.

7. What is the inductance of a coil that draws 0.25 A from a 110-V 60-Hz source? The resistance of the coil is 155 Ω.

8. Find the inductance of a coil that has a current changing at the rate of 125 mA/s. This current induces a counter emf of 35 mV.

9. Find the inductance of the coil whose current changes from 0.25 to 0.55 A in 1.25 s. The average voltage induced in the coil is 0.45 V.

JOBS 16–7 to 16–10   Transformers

Solve each of the following.

1. A No. 14 wire is used to wire a machine 35 m from a 60-Hz source. When the circuit is activated, the result is a 12.5-A current flow. Find the inductive reactance of the line.

2. Two No. 10 wires are used to supply 30 A to a machine located 110 ft from a 60-Hz power source. The spacing between the wires is 1.5 in. Find (*a*) the inductance, (*b*) the reactance, (*c*) the voltage drop, and (*d*) the Q of the cable.

3. A power transformer has a 300-V secondary. The primary has 600 turns and 115 V across it. Find the number of turns on the secondary winding.

4. Find the voltage delivered to the secondary of an auto transformer, if the primary has 700 turns, the secondary has 320 turns, and the primary voltage is 110 V.

5. A power transformer has a voltage ratio of 6 : 17. If the primary has 540 turns, find the number of turns on the secondary.

6. Find the turns ratio required to construct a 12-V bell transformer which is to operate on 115 Vac.

7. An electric soldering gun reduces the 115 V on the primary to 10 V on the secondary. Find the voltage ratio.

8. A power transformer whose primary is connected to a 110-V source delivers 15 V. If the secondary has 75 turns, find the number of turns on the primary. How many extra turns must be added to the secondary if it must deliver 60 V?

9. A soldering gun with a voltage ratio of 100 : 1 draws 1.5-A primary current. Find the secondary current.

10. A 12-V battery charger connected to a 110 Vac supplies a charging current of 15 A. Find the current in the primary.

11. A transformer with a 25 : 1 voltage step-down ratio has 9 V across 5 Ω in the secondary. Find (a) I_s, (b) I_p, and (c) P_s.

12. A power transformer delivers 73.5 W of power. If the primary voltage is 120 V, how much current is drawn by the primary?

13. The turns ratio of an auto transformer is 10 : 1. The primary voltage is 220 V. Find (a) V_s, (b) I_s, and (c) P_s. The secondary has a 45-Ω load across it.

14. A transformer is required to match a power transistor with a 12-Ω source resistance to a 4-Ω speaker. Find the turns ratio needed.

15. The secondary load of a step-down transformer with a turns ratio of 9 : 1 is 750 Ω. Find the impedance of the primary.

16. A transformer draws 155 mA from a 115-V line and delivers 55 mA at 450 V. Find its efficiency.

JOBS 16–12 and 16–13   Real Transformers

Solve each of the following.

1. A transformer draws 3 A at 110 V and delivers 9.5 A at 24 V. Find (*a*) the power input, (*b*) the power output, and (*c*) the efficiency.

2. A transformer draws 110 W and delivers 440 V at 150 mA. Find (*a*) the efficiency and (*b*) the primary current if the primary voltage is 110 V.

3. A power transformer that draws 750 W from a 220-V line operates at an efficiency of 85% and delivers 25 V. Find (*a*) the watts delivered and (*b*) the secondary current.

4. A transformer delivers 440 V at 75 mA at an efficiency of 90%. If the primary current is 825 mA, find (*a*) the power input and (*b*) the primary voltage.

5. A transformer delivers 486 V at 125 mA and at an efficiency of 90%. If the primary current is 750 mA, find (*a*) the power input and (*b*) the primary voltage.

JOB 16–14　　　　Real Cables and Power Distribution

Solve the following.

1. Using the equation $R = K \times \dfrac{L}{A}$, calculate the specific resistance for copper wire using the data

   for the following cable sizes: 14, 12, 10, 8.

Capacitance 17

| JOBS 17–1 to 17–6 | Capacitors |

Answer each of the following.

1. How does a capacitor react to alternating and direct current?

2. Name three factors that affect capacitance.

3. What is meant by *working voltage?*

4. Three capacitors, $C_1=20$ μF, $C_2=30$ μF, and $C_3=40$ μF, are connected in parallel. Determine the total circuit capacitance.

5. What are the total capacitance and working voltage of a 0.0035-μF 50-V capacitor, a 0.025-μF 100-V capacitor, and a 0.00065-μF 150-V capacitor when they are connected in parallel?

6. What is the total capacitance of the following capacitors: a 47 pF, a 0.000 072 μF, and a 0.000 000 05 μF?

7. Capacitors $C_1 = 20$ μF, $C_2 = 30$ μF, and $C_3 = 40$ μF are connected in series. What is the total capacitance?

8. A voltage divider supplying 440 V has two capacitors connected in series across it. The net capacitance is 330 μF. What is the proper value of each capacitor?

9. A series combination of 0.025- and 0.05-μF capacitors are connected in parallel with a 0.00215-μF capacitor. Find the total capacitance of the combination.

10. Two capacitors, 20 and 30 μF, are connected in series, and two other capacitors, 25 μF and 35 μF, are connected in parallel with the series combinations. Find the total capacitance.

11. Find the capacitive reactance of a 0.047-μF capacitor at (a) 15 kHz, (b) 105 kHz, and (c) 1 MHz.

12. Find the capacitive reactance of (a) 0.35-μF and (b) 0.00045-μF capacitors at a frequency of 60 Hz.

13. A 4-μF capacitor is connected to a 120-V 60-Hz source. Calculate (a) X_C and (b) I.

14. If the capacitance in question 13 were doubled, find the (a) X_C and (b) I.

15. A capacitor has a reactance of 7950 Ω at a frequency of 100 Hz. What is its capacitance?

16. A capacitance of 0.025 μF draws 0.02 A when connected across a 110-V source. What is the frequency of the ac voltage?

17. A capacitor produces a current of 20 A at 220 V when connected to a 60-Hz circuit. Find the value of the capacitor.

18. A 5.6-μF bypass capacitor passes a signal current of 150 mA at a frequency of 51 kHz. Find (a) the voltage drop across the capacitor and (b) the effective power consumed.

A Real Capacitor

Solve each of the following.

1. A 1200-Ω resistor and a 2-μF capacitor are connected in parallel across a 110-V 60-Hz source. Find (a) X_C, (b) I_T, (c) Z, (d) phase angle, (e) power factor, and (f) power drawn by the circuit.

2. A 10-kΩ resistor and a 0.01-μF capacitor are connected in parallel across a 120-V source. Find the effectiveness of the circuit by calculating the percent of the total current in the resistor for (a) a 1-kHz audio frequency and (b) a 10-MHz radio frequency.

3. A 300-Ω resistor and a 22.5-μF capacitor are connected in parallel across a power source of 108 V 60 Hz. Find (a) X_C, (b) I_R, (c) I_C, (d) I_T, (e) Z, (f) phase angle, (g) power factor, and (h) power drawn by the circuit.

4. A 1.5-kΩ resistor and a 4.2-μF capacitor are connected in parallel across a 220-V, 900-Hz source. Find the (a) X_C, (b) I_R, (c) I_C, (d) I_T, (e) Z, (f) phase angle, (g) power factor, and (h) true power.

Series AC Circuits **18**

c h a p t e r

JOBS 18–2 and 18–3	Capacitance in Series

Solve each of the following.

1. A resistor of 1.2 kΩ is connected in series with a capacitor that has a reactance of 2 kΩ. The current in the circuit is 75 mA. Find (a) Z, (b) V_R, (c) V_C, (d) V_T, and (e) the phase angle.

2. A resistor of 500 Ω is connected in series with an unknown capacitor, across a source of 115 V 60 Hz. The voltage drop across the resistor is 70 V. Find (a) I_T, (b) V_C, (c) phase angle, (d) capacitance of the capacitor, and (e) Z.

3. A circuit consisting of 200 Ω of resistance in series with a capacitor of 6.5 μF is connected across a source of 165 V 100 Hz. Find (a) X_C, (b) Z, (c) I_T, (d) V_R, (e) V_C, (f) phase angle, (g) power factor, and (h) true power.

4. A resistor of 85 Ω in series with a capacitor of 12.5 μF is connected across a 115-V source. What frequency will allow a current of 560 mA to flow?

5. A 55-μF capacitor in series with an unknown resistor is connected across a 120-V 60-Hz source. What value of resistance will allow a current of 250 mA to flow in the circuit?

6. A series-connected 0.25-μF capacitor, a 35-mH inductor, and a 150-Ω resistor are connected across a 12-V 3-kHz source. Find (a) X_C, (b) X_L, (c) Z, (d) I_T, (e) V_C, (f) V_L, (g) V_R, (h) phase angle, (i) power factor, and (j) true power.

7. An inductor of 425 mH with an internal resistance of 300 Ω in series with a capacitor of 12 μF is connected across a rectifier output to form a filter network. The rectifier output is 85 V at 60 Hz. Find (a) X_L, (b) X_C, (c) Q, (d) Z, (e) I_T, and (f) V_C. (If Q is greater than 5, we can neglect the resistance in computing Z.)

8. A resistor and capacitor connected in series across a 120-V 60-Hz source consume 85 W at a power factor of 93.9% leading. Find (a) I_T, (b) Z, (c) R, and (d) C.

JOBS 18–4 and 18–5	Series Resonance

Solve each of the following.

1. A resistor of 21 Ω, an inductance of 75 mH, and a capacitance of 0.23 μF are connected in series. Find (*a*) the resonant frequency, (*b*) the reactance of the inductance, (*c*) the reactance of the capacitance, (*d*) the impedance, (*e*) the current at resonance if the impressed voltage is 12 V, and (*f*) the voltage drop across the capacitance.

2. Find the value of a capacitor that must be connected across a 75-mH coil to make the circuit resonate at 60 Hz.

3. Find the resonant frequency of the series-resonant section of a bandpass filter that has an inductance of 157 μH and a capacitor of 13 μF.

4. Find what value of inductance must be placed in series with a 146-μF capacitor in order to provide resonance for a signal of 610 Hz.

5. A 235-mH inductance, a 45-Ω resistance, and a 0.15-μF capacitance are connected in series. Find (*a*) the impedance of the circuit to a frequency of 50 Hz, (*b*) the capacitance that must be added in parallel with the 0.15-μF capacitor to produce resonance at this frequency, and (*c*) the impedance of the circuit at resonance.

Parallel AC Circuits 19

JOBS 19–1 to 19–4   Parallel AC Circuits

Solve each of the following.

1. A 75-Ω incandescent lamp and a 20-Ω electric iron are connected in parallel across a 115-V 60-Hz power source. Find (a) the total resistance, (b) the total current, (c) the power factor, and (d) the power consumed.

2. Two inductances, of 6 and 9H, are connected in parallel across a 110-V 60-Hz power source. Find (a) the current in each inductance, (b) the total current, (c) the impedance, (d) the power factor, and (e) the power consumed.

3. Two pure capacitances of 0.032 and 0.52 μF are connected in parallel across a 105-V 60-Hz line. Find (a) the current in each capacitance, (b) the total current, (c) the impedance, (d) the power factor, and (e) the power consumed.

4. An inductor of 265 mH and a resistor of 6000 Ω are connected in parallel to form a high-pass filter circuit. They are connected across a 115-V source. Find the effectiveness of the circuit by calculating (a) the branch currents, (b) the total current, and (c) the percent of the total current in the resistor for a 1-kHz audio frequency and 1-MHz radio frequency.

5. A 35-Ω resistor, a 45-Ω inductive reactance, and a 24-Ω capacitive reactance are connected in parallel across a 115-V 60-Hz power source. Find (*a*) branch currents, (*b*) total circuit current, (*c*) phase angle, (*d*) impedance, (*e*) power factor, and (*f*) true power drawn by the circuit.

6. Find the equivalent series circuit to the circuit in question 5.

7. A 30-Ω resistor, a 60-Ω inductive reactance, and an 80-Ω capacitive reactance are connected in parallel across a 110-V 60-Hz source. Find (*a*) I_R, (*b*) I_L, (c) I_C, (*d*) I_T, (*e*) phase angle, (*f*) impedance, (*g*) power factor, and (*h*) power drawn by the circuit.

8. A current of 12.9 A in one branch of an ac circuit leads the total voltage by 60°. Find its in-phase current and its reactive current.

9. A current of 4.5 A supplying an induction motor lags the voltage by 37°. Find the in-phase and reactive currents.

JOB 19-6   Parallel-Series AC Circuits

Draw a typical circuit and response curve for each of the following.

1. Band-pass (passes a band of frequencies to the load).

2. Band-pass (diverts a band of frequencies away from the load and passes the required frequencies to the load).

3. Band-stop (blocks a band of frequencies from reaching the load).

4. Band-stop (diverts a required band of frequencies from reaching the load).

5. High-pass (blocks frequencies below a certain cut-off frequency).

6. Low-pass (blocks frequencies above a certain cut-off frequency).

Solve each of the following.

7. A synchronous motor of 30 Ω resistance and 60 Ω capacitive reactance is in parallel with an induction motor of 24 Ω resistance and 32 Ω inductive reactance and a third parallel branch of 20 Ω resistance. Find (*a*) the total current drawn from a 240-V 60-Hz source, (*b*) the total impedance, (*c*) the phase angle, (*d*) the power factor, and (*e*) the power drawn by the circuit.

8. Two impedances are connected in parallel. One branch has a resistor of 17.32 Ω in series with an inductance that has an inductive reactance of 10 Ω. The second branch has a resistor of 10 Ω in series with a capacitor that has a capacitive reactance of 17.32 Ω. The two branches are connected across a 240-V 60-Hz source. Find (*a*) the total current, (*b*) the circuit impedance, (*c*) the power drawn by the circuit, (*d*) the phase angle, and (*e*) the power factor.

JOB 19–7   Series-Parallel AC Circuits

Solve the following.

1. Solve the circuit shown in Fig. 19-1 for (*a*) the equivalent series impedance, (*b*) the total current, (*c*) the phase angle, (*d*) the power factor, and (*e*) the power drawn by the circuit.

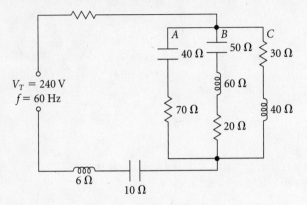

$V_T = 240$ V
$f = 60$ Hz

Figure 19-1

JOB 19-8	Parallel Resonance

Solve each of the following.

1. A parallel resonant tank circuit consists of a coil with an inductance of 230 μH and a capacitor of 75 pF. What is the resonant frequency that the circuit will reject?

2. A 16-μH coil and a variable capacitor are connected in parallel to form a tuned circuit. If the circuit is to be resonant to 5622 kHz, what must be the value of the capacitor?

3. An IF coil in a superheterodyne receiver resonates at a frequency of 1624 kHz. Find the inductance of the coil if the capacitor has 120 pF.

Alternating-Current Power

20

c h a p t e r

Power and Power Factor

Solve each of the following.

1. An electric motor draws 20 A from a 120-V 60-Hz source. It consumes 2.2 kW. Find the power factor.

2. A resistor of 22 Ω resistance and a capacitor of 15 Ω reactance is connected to a 120-V 60-Hz ac line. Find (*a*) the power factor and (*b*) the effective power.

3. Find the true power used by a synchronous motor operating at a power factor of 87% if it draws 5.6 A at 110 V.

4. An induction motor connected to a 220-V 60-Hz source has an effective inductance of 2 H and its resistance is 425 Ω. Find (*a*) the power factor, (*b*) the true power, and (*c*) the current drawn by the motor.

5. A motor operating at 75% power factor consumes 247 W from a 440-V 60-Hz line. Find the current flow in the circuit.

6. The electrical load in a small shop consumes 1.86 kW of power. The power factor is 65%. Find the voltamperes of power delivered to the shop.

7. An induction motor takes 75 kVA at a power factor of 85%. The efficiency is 90%. Find the power delivered to the motor.

JOB 20–2   Total Power Drawn by Combinations of Reactive Loads

Solve each of the following.

1. One motor takes 60 kVA at 50% power factor lagging, and another motor connected to the same power line takes 110 kVA at 85% power factor lagging. Find (a) the total effective power, (b) the total reactive power, (c) the total PF, and (d) the total apparent power.

2. An induction motor draws 21 kW at a power factor of 60%, and another induction motor draws 43 kW at 75% power factor. Find (a) the PF of the line, (b) the total apparent power, and (c) the total effective power.

3. A 20-hp induction motor at 75% PF is connected to the same line as a 45-hp induction motor at 90% PF. Find (a) the total apparent power taken by the motors and (b) the total PF.

| JOB 20–3 | Power Drawn by Combinations of Resistive, Inductive, and Capacitive Loads |

Solve each of the following.

1. A 25-kVA induction motor at an 80% lagging PF and a synchronous motor that takes 40 kVA at a 60% leading PF are connected to an alternator. Both motors are connected in parallel across a 220-V 60-Hz ac line. Find (*a*) the total effective power, (*b*) the total PF, (*c*) the total apparent power, and (*d*) the total current drawn.

2. Three loads—a 15 A at unity power factor, a 35 A at a power factor of 0.6 lagging, a 25 A at a leading power factor of 0.8—are connected across a 220-V source. Find (*a*) the total power, (*b*) voltamperes, (*c*) the reactive volt-amperes, (*d*) the overall power factor, and (*e*) the total current.

3. An industrial system uses two induction motors. One motor is rated at 30 A and 0.80 lagging power factor, and the second motor is rated at 40 A and 0.85 power factor. The supply voltage to the system is 240 V. A synchronous motor, rated at 25 A and 0.65 leading power factor, is used to improve the overall system power factor. Find (*a*) the true, apparent, and reactive power for each of the two induction motors, (*b*) with only the two induction motors on line, find the total true, apparent, and reactive power supplied to the two induction motors, (*c*) the power factor with only the two induction motors in operation, (*d*) the true, apparent, and reactive power for the synchronous motor, (*e*) the total system true, apparent, and reactive power when all three motors are in operation, (*f*) the overall system power factor, and (*g*) the supply current with all three motors in operation.

4. The electrical load of an industrial system is 800 kVA at a power factor of 0.75 lag. The system is operated from a 4800-V supply. A synchronous motor is to be added to the system to carry an additional load of 200 kW with a 0.85 leading power factor. Find (*a*) the true reactive power of the original motor, (*b*) the power factor after the synchronous motor has been added to the system, and (*c*) the total current with both motors operating in the system.

JOB 20-4   Power-Factor Correction

Solve each of the following.

1. An induction motor takes 800 kVA at 2400 V and has a power factor of 70% lag. What must be the PF of a 900-kVA synchronous motor connected in parallel in order to raise the total PF to 100%?

2. A 54-kW 0.75 power factor load is connected to an ac line. Find (*a*) the apparent power, the (*b*) phase angle, (*c*) the lagging reactive power, (*d*) the number of standard 15- or 25-kvar capacitors needed to improve the PF, and (*e*) the new PF.

3. An industrial system has a load of 66 kW with a power factor of 65%. How much capacitor kvar is required to raise the power factor to 95%? How many standard 25-kvar capacitors would be needed?

The Wye Connection **21**

JOB 21-1	The Wye Connection

Solve each of the following.

1. List four advantages of three-phase systems over single-phase systems.

2. If the phase voltage of a wye-connected generator is 440 V, find the line voltage.

3. If the line voltage of a wye system is 230 V, find the voltage across the coil of the generator.

4. The phase current of a wye-connected generator is 80 A. Find the line current.

5. A three-phase load draws a current of 45 A per terminal. The line voltage is 220 V, and the power factor is 85%. Find the power consumed by the load.

6. A wye-connected generator produces a phase current of 45 A at a phase voltage of 1386 V and a power factor of 90%. Find (a) the generator line voltage, (b) the power produced by each phase, and (c) the total three-phase power.

7. A three-phase load consumes 75 kW from a 550-V line. The line current is 95 A. Find (a) the kilovoltamperes and (b) the power factor of the load.

8. A three-phase wye-connected generator is connected to an industrial motor load. The generator has an output coil voltage of 120 V, a coil current of 15 A, and a lagging power factor angle of 40°. Find (a) the line voltage, (b) the line current, (c) the total apparent power, and (d) the total true power.

9. The voltage across each phase of a wye-connected source is 230 V. What will be the current in each of the three line wires if a balanced load of 175 kW at a power factor of 0.85 is delivered?

10. A wye-connected load of three 75-Ω resistors is connected to a 250-V three-phase supply. The power factor is 100%. Find (a) the voltage across each resistor, (b) the current in each resistor, (c) the line current, (d) the total power used, and (e) the total kilovoltampere input to the load.

11. A 440-V three-phase source serves a balanced wye-connected load consisting of three equal impedances. Each load consists of a 4.5-Ω resistor in series with an inductive reactance of 5.8 Ω. Find (*a*) the impedance per phase, (*b*) the power factor, (*c*) the total current in each branch of the load, (*d*) the line current, (*e*) the total power, and (*f*) the kilovoltampere input.

12. A 10-hp 220-V three-phase wye-connected motor draws a current of 30 A per terminal when operating at rated output. Find (*a*) the full-load kilovoltampere input and (*b*) the kilowatt input if the full-load power factor is 90%.

JOB 21-2   The Delta Connection

Solve each of the following.

1. The current through each phase of a delta-connected generator is 45 A. Find the line current.

2. A delta-connected three-phase generator is connected to an industrial network. The generator has a rated phase voltage of 8000 V and a rated current of 425 A. At full load, its power factor is 80% lag. Find (*a*) the line voltage, (*b*) the line current, (*c*) the generator kVA rating, and (*d*) the full-load power in kilowatts.

3. A three-phase delta-connected industrial motor load is supplied with a three-phase line voltage of 480 V and draws a line current of 20 A. If the load power factor is 70% lag, determine (*a*) the circuit phase voltage, (*b*) the phase current, (*c*) the load apparent power, and (*d*) the load true power.

4. A balanced delta-connected three-phase load has a phase resistance $R_P = 12 \ \Omega$ in series with an inductive reactance $X_P = 8 \ \Omega$. If the phase voltage to the load is equal to 360 V, determine (*a*) the phase impedance, (*b*) the load power factor, (*c*) the phase current, (*d*) the line current, and (*e*) the load true power.